Technologies for RF Systems

For a listing of recent titles in the
Artech House Microwave Library,
turn to the back of this book.

Technologies for RF Systems

Terry Edwards

ARTECH
HOUSE

BOSTON | LONDON
artechhouse.com

Library of Congress Cataloging-in-Publication Data
A catalog record for this book is available from the U.S. Library of Congress.

British Library Cataloguing in Publication Data
A catalog record for this book is available from the British Library.

ISBN-13: 978-1-63081-450-2

Cover design by John Gomes

© 2018 Artech House

All rights reserved. Printed and bound in the United States of America. No part of this book may be reproduced or utilized in any form or by any means, electronic or mechanical, including photocopying, recording, or by any information storage and retrieval system, without permission in writing from the publisher.

All terms mentioned in this book that are known to be trademarks or service marks have been appropriately capitalized. Artech House cannot attest to the accuracy of this information. Use of a term in this book should not be regarded as affecting the validity of any trademark or service mark.

10 9 8 7 6 5 4 3 2 1

For Patricia

Contents

Acknowledgments   xv

CHAPTER 1
Twenty-First Century RF Systems and Electronics   1

1.1	Introduction	1
1.2	Abbreviations Relating to Symbols Used in this Book	3
1.3	Antennas	3
1.4	The Challenge of Frequency Bands and Wavelengths	6
1.5	Software-Defined Radio and Cognitive Radio	9
1.6	The Challenge of Noise	10
1.7	RF Receivers	11
1.8	RF Filters	12
1.9	ADCs and DACs	14
1.10	Oscillators, Mixers, and Frequency Converters	14
1.11	Semiconductor Device Requirements	17
1.12	Semiconductor Manufacturing	17
1.13	Diodes and Transistors	18
1.14	Hybrid Circuits and MMICs	19
1.15	The Challenge of RF Power Amplification	20
1.16	Electronic Design Automation	21
	References	22

CHAPTER 2
RF Semiconductors   23

2.1	Introduction	23
2.2	Semiconductor Materials	23
	2.2.1 Bandgap	24
	2.2.2 Drift Velocity	25
	2.2.3 Resistors Made from Semiconductors	26
	2.2.4 Electron Speed and Transit Time	27
	2.2.5 Some Further Important Properties of Semiconductors	27
	2.2.6 Semiconductor Manufacturing	28
2.3	Semiconductor Diodes (RF-Oriented)	28
	2.3.1 Some Semiconductor Junction Diode Fundamentals	28
	2.3.2 P-I-N Diodes	30

		2.3.3 Varactor Diodes	31
		2.3.4 Noise in RF Diodes	32
	2.4	Transistors	33
		2.4.1 Introductory Remarks	33
		2.4.2 High Frequency Circuit Models for Transistors	34
		2.4.3 CMOS and Related Transistor Technologies	35
		2.4.4 GaAs and GaN Field-Effect Transistors	36
		2.4.5 The GaAs HEMT and pHEMT	37
		2.4.6 The GaN HEMT	38
		2.4.7 Bipolar RF Transistors	39
	2.5	MMICs and RFICs	41
		References	42

CHAPTER 3

Passive RF Components — 43

3.1	Introduction	43
3.2	Discrete Passive RF Components	43
	3.2.1 Capacitors	43
	3.2.2 Inductors	45
	3.2.3 Resistors	47
3.3	RF Transmission Lines	48
	3.3.1 Coaxial Lines	49
	3.3.2 Microstrip	50
3.4	Coplanar Waveguide	60
3.5	Substrate Integrated Waveguide	61
	References	62

CHAPTER 4

Passive RF Circuit Elements — 63

4.1	Introduction	63
4.2	Fundamentals of Directional Couplers	63
4.3	The Lange Coupler	64
	4.3.1 EM Structure	66
4.4	Wilkinson Power Dividers	67
	4.4.1 Introduction to Wilkinson Dividers	67
	4.4.2 Equal-Split Wilkinson Dividers	67
	4.4.3 Unequal-Split Wilkinson Dividers	68
	4.4.4 Multiport Equal-Split Wilkinson Dividers	70
4.5	Baluns	72
	References	74

CHAPTER 5

Switches, Attenuators, and Digital Circuits — 75

5.1	Introduction	75
5.2	Solid State RF Switches	75
	5.2.1 Some Overall Aspects	75

	5.2.2	Reflective and Nonreflective SPDT GaAs FET Switches	76
5.3	Attenuators		78
5.4	Digital Circuits		80
	5.4.1	Selected Examples of Logic Gates	80
	5.4.2	Digital Signal Processors	81
	5.4.3	Electronically Programmable Read-Only Memories	82
	5.4.4	Field-Programmable Gate Arrays	83
	5.4.5	Provision for Built-In Test and Related Requirements	84
	5.4.6	Technology Utilized for Digital Circuit Elements	84
		References	85

CHAPTER 6

Radio-Frequency Filters — 87

6.1	Introduction	87
6.2	Review of Basic Concepts and Fundamentals	87
6.3	Technology Options	89
6.4	LPFs Formed with Cascaded Microstrips	90
6.5	Microwave BPFs	92
6.6	Suspended Substrate Stripline Filters	96
6.7	Inline Microstrip Filter Structures	97
6.8	Filters Using Defected Ground Plane Technology	98
6.9	Dielectric Resonators and Filters Implementing Them	98
6.10	SIW-Based BPFs	100
6.11	Millimeter-Wave BPFs	101
6.12	Tunable BPFs	102
	References	102

CHAPTER 7

Antennas — 105

7.1	Introduction		105
7.2	Antenna Fundamentals		106
	7.2.1	Near-Field and Far-Field Conditions	107
	7.2.2	Radiation Patterns and Beamwidth	108
	7.2.3	Directivity	109
	7.2.4	Radiation Efficiency	109
	7.2.5	Aperture Efficiency	110
	7.2.6	Effective Area	111
	7.2.7	Gain	111
	7.2.8	Equivalent Isotropic Radiated Power	112
	7.2.9	Friis' Equation	112
	7.2.10	Impedance Matching	113
	7.2.11	Polarization	113
	7.2.12	Antenna Noise Temperature	114
	7.2.13	Gain-Temperature Ratio	115
7.3	Dish Reflector Antennas		116
7.4	Flat-Panel or Patch Antennas		117

7.5	Analog, Digital, and Hybrid Beamforming	118
7.6	Active Electronically-Scanned Arrays	119
	References	121

CHAPTER 8

Small-Signal RF Amplifiers — 123

8.1	Review of Amplifier Fundamentals	123
8.2	Basic RF Amplifiers	125
	8.2.1 Practical RF Amplifier Realization	125
	8.2.2 Interstage or Inner Matching Networks	126
8.3	The Vital Issue of Stability	127
8.4	Fundamental Receiver Characteristics Leading to the Need for AGC	129
	8.4.1 Toward an Effective AGC Circuit Design	129
8.5	High-Gain RF Amplifiers	131
8.6	Broadband Amplifiers	134
	8.6.1 Basic Requirements	134
	8.6.2 Balanced Amplifiers	135
	8.6.3 Distributed Amplifiers	136
	References	138

CHAPTER 9

Noise and LNAs — 141

9.1	Introduction	141
9.2	Noise Factor, Noise Figure, and Equivalent Noise Temperature	142
9.3	Noise Figure for an Attenuating Element	144
9.4	Minimum Detectable Signal	145
9.5	Noise in Transistors	146
	9.5.1 Thermal Noise, Particularly Thermal Diffusion Noise	147
	9.5.2 Shot Noise	148
	9.5.3 Flicker Noise	148
	9.5.4 Phase Noise	149
	9.5.5 Variation of Noise Figure with Frequency	150
9.6	Overall Noise Figure for Cascaded Blocks	151
9.7	Noise-Matching and Narrowband LNA Design	156
	References	159

CHAPTER 10

RF Power Amplifiers — 161

10.1	Introduction	161
10.2	Some Basic Aspects of RFPAs	161
10.3	Transistor Choices, Hybrid Circuits, and MMICs	162
10.4	Power Levels, Power Gains, and Efficiency	163
	10.4.1 Internal Transistor Output Characteristics	163
	10.4.2 RFPA Output-Input Power Transfer Characteristics	164
	10.4.3 Amplifier Efficiency	164

10.5	Compression and Peak-to-Average Power Ratio	166
	10.5.1 Compression and a Summary of Main Parameters	166
	10.5.2 Peak-to-Average Power Ratio	167
10.6	Error Vector Magnitude	167
10.7	Classifications of Power Amplifiers	168
	10.7.1 Class A Amplifiers	168
	10.7.2 Class B and AB Amplifiers	170
	10.7.3 Class C Amplifiers	171
10.8	Harmonically Matched Power Amplifiers	171
	10.8.1 Switched-Mode RFPAs	171
	10.8.2 Class F Power Amplifiers	175
10.9	The Doherty Power Amplifier Configuration	178
10.10	The Envelope-Tracking Amplifier	180
10.11	High Power Push-Pull Amplifiers	181
10.12	Other Practical RFPA Circuits	181
	10.12.1 Ka-Band PA MMIC Examples	182
10.13	The Distortion Issue and Linearization Techniques	183
	10.13.1 Linearity and Intermodulation Distortion	183
	10.13.2 Linearization Techniques	185
10.14	Some Final Overall Comments Regarding RFPAs	186
	References	187

CHAPTER 11

RF-Oriented ADCs and DACs — 189

11.1	Introduction	189
11.2	ADCs	189
	11.2.1 Quantization and Sampling	189
	11.2.2 Sampling in Practical ADCs	191
	11.2.3 Effective Number of Bits	191
	11.2.4 Quantization Error and Quantization Noise	193
	11.2.5 Quantization Static Error and Sampling Distortion	194
	11.2.6 Sampling Jitter	195
	11.2.7 Aliasing and Antialiasing	197
	11.2.8 Adjacent Channel Power Ratio	199
11.3	ADC Architectures	200
	11.3.1 The Flash ADC Architecture	200
	11.3.2 The Folding ADC Architecture	201
	11.3.3 Pipelined ADC Architecture	201
	11.3.4 Time-Interleaved ADCs	202
11.4	Digital-to-Analog Converters	204
	11.4.1 Basic Structure and Functionality of a DAC	204
	11.4.2 DAC Resolution, Speed, and Figures of Merit	204
	11.4.3 Some Practical Aspects of High-Speed DACs	207
	References	207

CHAPTER 12

Radio Frequency Sources — 209

12.1 Some Fundamental Aspects of RF Oscillators — 209
12.2 Quartz Crystal Oscillators — 210
    12.2.1 The Quartz Crystal — 210
    12.2.2 Quartz Crystal-Based Oscillators — 211
12.3 Oscillators Controlled by Dielectric Resonators — 212
12.4 VCOs — 214
12.5 Importance and Impact of Phase Noise — 215
12.6 Frequency Multipliers — 220
12.7 Frequency Dividers — 221
12.8 Phase-Locked-Loop-Based Frequency Synthesizers — 222
    12.8.1 Basic Configuration — 222
    12.8.2 The Fractional-N Frequency Synthesiser — 222
    References — 224

CHAPTER 13

Frequency-Band Conversion — 225

13.1 Introduction — 225
13.2 Fundamentals of Mixers — 226
    13.2.1 Basic Features — 226
    13.2.2 Image Frequency — 227
13.3 Diode-Based Mixers — 228
    13.3.1 The Single-Ended Diode Mixer — 228
    13.3.2 The Double-Diode Mixer — 230
    13.3.3 The Image-Reject Mixer — 231
    13.3.4 Upconverters — 232
13.4 Transistor-Based Mixers — 233
    13.4.1 The Single-Ended FET Mixer — 233
    13.4.2 Differential FET Mixer — 234
    13.4.3 CMOS-Based Mixers — 235
    13.4.4 Mixer Implementing a Cascode Circuit — 236
    13.4.5 The Gilbert Cell Mixer — 236
    References — 239

CHAPTER 14

Modulation Techniques and Technologies — 241

14.1 Introduction — 241
14.2 Amplitude Modulation — 242
14.3 Frequency Modulation — 245
14.4 Digital Modulation — 247
    14.4.1 Specific Aspects Relating to Digitally Modulated Systems — 247
    14.4.2 ASK, OOK, and FSK — 250
    14.4.3 BPSK and QPSK — 251
    14.4.4 M-PSK, QAM, and APSK — 255

	14.4.5	Spectral Efficiency of the Various Digital Systems	257
	14.4.6	Probability of Bit Error or Bit Error Rates	257
	14.4.7	Closed-Form Expressions for the Complementary Error Function	259
	14.4.8	BER Data Compared	259
	14.4.9	Spread-Spectrum Modulation	260
	14.4.10	Orthogonal Frequency Division Multiple Access	262
14.5	Transceivers		262
	14.5.1	Basic Concept of a Transceiver	262
	14.5.2	Software-Defined Radio	263
	14.5.3	Full-Duplex Radios	263
	14.5.4	Transceiver Modules for Short-Range Radio	263
		References	264

Appendix A
Logarithmic Units — 265

Appendix B
S-Parameters and X-Parameters — 269

B.1 Scattering (S)-Parameters — 269
B.2 X-Parameters — 270
    References — 272

Acronyms and Abbreviations — 273

About the Author — 277

Index — 279

Acknowledgments

In writing any book, one requires inspiration, passion, perseverance, time, knowledge (including 'where to find things'), and, above all these aspects, the invaluable support of others. Easily my mainstay has come from my wife Patricia Adene, without whose support and patience this book would never have been completed.

There are several other people whom I want to *sincerely* thank. Among them are:

- Nick Riley, who was my manager at the University of Hull, where I provided part-time lectures on RF/microwave technology for the communications systems M.Sc. course. Nick initiated the idea I might write this book.
- Don Black and Dave Taunton, whom I taught and who then graduated from La Trobe University, Melbourne, Australia, with B. Comm. (Eng.) degrees. Don and Dave were the first people to suggest to me that I should write a book.
- Malcolm Edwards and Andrew Wallace of NI AWR, who were greatly helpful in terms of providing access to EDA examples.
- Steve Edwards, a good friend who so ably and professionally prepared approximately 90% of all the drawings in this book.
- Brandon Browne, who, at the age of 16, led me to the desmos program and hence the generation of representative amplitude and frequency modulation waveforms.
- My reviewer, without whom many calculation results in this book would have been inaccurate and much important material would have been missed. My reviewer has also excellently served to encourage me during the sometimes quite stressful activities toward the completion of this work.
- Oren Hagai, CEO of Interlligent, for his support in the area of signal converters
- Finally, Steve Manton, who expertly rechecked the math.

CHAPTER 1

Twenty-First Century RF Systems and Electronics

1.1 Introduction

For well over a century, radio frequency (RF) technology has been understood in sufficient detail for the design of basic communications systems. The application of RF to radar began in the 1930s and accelerated during World War II, driven by the pressing needs of the major war effort.

In both instances, communications and radar, electronic vacuum tubes (or valves) dominated the scene regarding almost all requirements relating to active devices. What changed the entire electronics scene forever were the pivotal inventions of the transistor in 1947 and the integrated circuit (IC) in 1958. Key developments in microwave integrated circuits during the 1950s (right up to the present era) ensured an ever-advancing solid state era for RF technology. Twenty-first century RF systems exhibit the following trends: they are increasingly digital, increasingly software-based, and almost entirely solid state.

The overall physical dimensions of critical components and devices involved in communications systems and radars embrace an extremely wide range from massive communications towers (and phased-array radars) all the way down to the nanometer-scale semiconductor devices involved in the electronics (see Figure 1.1). Between these extremes, there exist many types of modules and subsystems that perform specific signal-processing functions, most of which are described in detail within this book.

Monolithic microwave integrated circuits (MMICs) typically embody several transistors on the same chip, some tens through several hundred in the case of relatively complex silicon realizations.

If all the transistors (mainly digital ICs and MMICs) involved all the subsystems involved in the Hamburg Tower were counted up the total would amount to many billions. And the great majority will be taken up by the highly transistor-intensive digital ICs.

A technique called space-division multiple access (SDMA) is used extensively as part of the infrastructure for mobile (cellular) networks. This arrangement is illustrated conceptually in Figure 1.2.

(a)

(b)

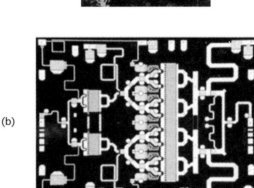

(c)

Figure 1.1 Physical scales of RF communications technologies (in descending orders of magnitude). (a) A photograph of the Hamburg Tower in Germany. (Horizon House Publications are thanked for permission to use this image, a photograph originally taken by Kristof Hamann, in [1].) (b) A MMIC chip capable of 10W of output microwave power. (Transcom, Inc., are thanked for their permission to reproduce this image.) (c) A small snowflake shown approximately to scale.

The narrow electronically steerable beam picks out the required user for the (very short) periods of time required in each instance. The entire system comprises a highly dynamic and extensively integrated network-of-networks.

1.2 Abbreviations Relating to Symbols Used in this Book

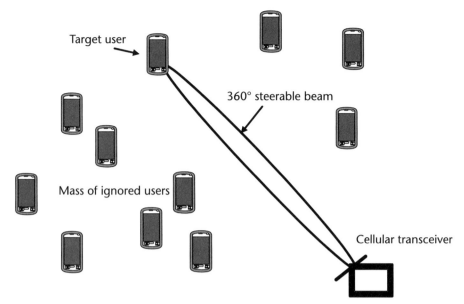

Figure 1.2 Part of a cellular network indicating a steerable beam for SDMA. (In reality, the beam does not shrink with distance as shown here; actually it expands. The focused shrinkage just indicates the concentration of the single target user.)

A major trend in mobile communications networks involves the development and application of small cells and what is termed densification. This will mean far denser subnetworks than hitherto conceived, mainly around urban regions, globally. It also means quite a dramatic shift toward millimeter-wave links and associated technology in order to obtain the much higher bit rates (typically several gigabits per second) that are greatly desired.

The remaining sections of this chapter start with presenting a summary of the symbols used through the book. Following this there are brief descriptions of current and prospectively important antennas and systems. Finally some significant subsystems are described before proceeding all the way down to the devices and components level.

1.2 Abbreviations Relating to Symbols Used in this Book

Most of the component and circuit symbols used throughout this book are fairly standard and a summary is presented in Table 1.1.

The only device not included in Table 1.1 is the bipolar junction transistor, shown in Figure 1.3.

1.3 Antennas

Transmitting and receiving antennas are often the clearest external evidence of an RF system (for example, the total number of microwave and millimeter-wave dish

Table 1.1 Common RF Components with Descriptions and Symbols

Component	Description	Symbol
Absorber	Material (generally ferrite-loaded) that absorbs electromagnetic energy	
Antenna	Radiates electromagnetic energy into free space	
Attenuator	Resistive element that adds loss	
Balun	Transformer that converts a balanced signal (two signals with no fixed ground) to an unbalanced signal (clear common ground)	
Bias tee	Three-port network that combines DC with RF or takes a signal and splits it into DC and RF components	
Capacitor	Basic element that blocks DC and passes AC and that can also store charge	
Circulator	Three- (or more) port network that restricts the flow of electromagnetic energy to one direction	
Coupler	Four-port network that splits the input to two equal or unequal amplitude outputs and that has an isolation port	
Diode	Basic element that only passes current in one direction (the direction the triangle points)	
Diplexer	Three-port network that splits into two ports with different frequency responses	
Duplexer	Allows a transmitter and receiver to share a single antenna	
Equalizer	Flattens a response (such as gain) over frequency	
Filter	Changes the amplitude of a signal based on the frequency response (band pass filter shown)	
Inductor	Basic element that blocks AC and passes DC and stores magnetic flux	
Isolator	Two-port network that restricts the flow of electromagnetic energy to one direction	
Limiter	Prevents output power from exceeding a threshold	
Low-noise amplifier (LNA)	Amplifier optimized for high gain and low noise generation	
Mixer (downconverter)	Multiplies an input signal (RF) by a fixed frequency (LO) to downconvert to an intermediate frequency (IF)	

1.3 Antennas

Table 1.2 (continued)

Component	Description	Symbol
Power amplifier	Amplifier optimized for high output power	
Power combiner	Multiport network that combines multiple input ports into a single output port with increased amplitude	
Power splitter	Multiport network that splits a single input into multiple output ports with reduced amplitude	
Resistor	Basic element that attenuates voltage	
Switch	Basic element that directs a signal from one path to another	
Thermistor	Resistor with predictable temperature response	
Transistor	Voltage-controlled resistor and basic element in an amplifier	
Varactor	Voltage or mechanically tunable capacitor	

Source: [2]

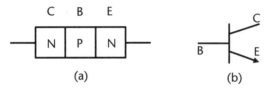

Figure 1.3 Schematic structure (a) and circuit symbol (b) for an NPN BJT.

reflector antennas that are fitted on the Hamburg Tower shown in Figure 1.1[a] approaches 100).

But dish and other large-scale antennas are only a part of the radiating elements story. In situations where substantial antenna gain is not a priority, it is possible to design and implement planar antennas and an example is shown in Figure 1.4 [2].

In this structure, metamaterials and composite right- or left-handed transmission lines (CRLH-TLs) are implemented. A good reference regarding the principles and applications of metamaterials has been provided by Brookner [3].

Flat-panel antenna arrays are increasingly being implemented into the multiple input/multiple output (MIMO) systems that are being put forward as potential candidates for several new types of communications systems, notably the 5G. A technique known as beamforming is necessary to shape and direct the beams associated with the flat-panel arrays and either analog or digital beamforming can be applied. Hybrid beamforming combines the analog and digital approaches. The article by Amitava Ghosh [4] provides an excellent overview of the technology.

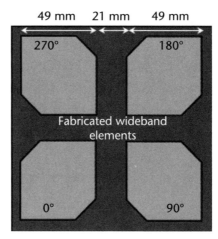

Figure 1.4 Four-element wideband patch antenna array. (Based on a photograph in [5].)

1.4 The Challenge of Frequency Bands and Wavelengths

The unwary engineer might just possibly be forgiven for assuming anyone, any engineer, can simply choose some likely practical-looking frequency band and design and manufacture the required circuit.

However, this approach would, if unfettered, lead to total spectral anarchy and instead a disciplined approach has been instituted, principally via the International Telecommunication Union (ITU). Table 1.2 exhibits the main (RF) frequency bands designated by the ITU.

There are several further frequency bands accommodating signals below 30 kHz but the RF bands are all cited in Table 1.2.

Microwave bands mainly encompass ultrahigh frequency and superhigh frequency. Millimeter-wave bands (strictly speaking) are within extrahigh frequency. However, it is notable that extrahigh frequency covers a massive 270-GHz range of frequencies. Millimeter-wave bands extend to well over 100 GHz, after which the term submillimeter becomes more widely used (i.e., corresponding to free-space wavelengths below 1 mm).

The term radio frequency applies in general to electromagnetic signals operating at frequencies ranging from around 30 kHz through to over 100 GHz. The fundamental unit of frequency is the hertz (Hz), which means that 1 kHz equals 10^3 Hz and 1 GHz equals 10^9 Hz.

Carrier frequencies ranging from 0.5 GHz to 6 GHz are very important for terrestrial systems (including cellular mobile), while 12.4 to 18 GHz, known as Ku-band, remains in substantial use for many satellite communications systems.

Table 1.2 Main RF Communications Frequency Bands

Band Name	Low Frequency	Medium Frequency	High Frequency	Very-High Frequency	Ultrahigh Frequency	Superhigh Frequency	Extrahigh Frequency
Frequency Range	30–300	0.3–3	3–30	30–300	0.3–3	3–30	30–300
Units	kHz	MHz	MHz	MHz	GHz	GHz	GHz

1.4 The Challenge of Frequency Bands and Wavelengths

Meanwhile, millimeter-wave (in practice around 26 GHz and above) is increasingly used for satellite systems and also prospectively for terrestrial communications' 5G links.

Historically originating in radar systems (with rectangular waveguide technology) but in general use in most microwave and millimeter-wave systems, there are the specific two groups of letter-designated bands shown in Table 1.3.

Unfortunately, each sequence of bands shown in Table 1.3 shares two letters: I and L—although the actual frequency bands differ markedly. As usual, careful questioning is needed in specific situations to decide upon precisely which frequencies are intended. The NATO bands are rarely if ever encountered outside of certain military (especially European) scenarios.

It follows, for example, that when a system or a part of a system is described as operating within the X-band it could be utilizing the entire 8.2 to 12.4-GHz band or a portion thereof (e.g., 9 to 10 GHz).

The terms microwave and millimeter wave are commonly used and it is important to clarify these terms as closely as possible. In general, microwave refers to signals ranging from around 500 MHz (middle of the ultrahigh frequency) to 26 GHz, which is almost the top of the K-band. In practice, millimeter waves extend from 26 GHz to over 100 GHz. Very often, it is stated that millimeter waves start at 30 GHz, although this is really a theoretical approximation and in practice 26 GHz is commonly understood. This is important, for example, in view of the first officially approved millimeter-wave 5G frequency band, which is in the 26–29.5-GHz range (indicating a 3.5-GHz operating bandwidth).

Higher-frequency millimeter-wave bands such as V-band and E-band are also important. The definitions are:

- V-band: 57 to 64 GHz;
- Lower E-band: 71 to 76 GHz;
- Upper E-band: 81 to 86 GHz.

Note that the V-band is 7 GHz wide and is principally focused on the often-used 60-GHz near-center frequency. There is a gap of 7 GHz between the upper extreme of the V-band and the beginning of the lower E-band. The lower and upper E-bands both have bandwidths of 5 GHz. The gap between the lower and upper E-bands is reserved for the nominally 77 GHz automotive (ACC) radars.

Table 1.3 Standard Frequency Band (Letter) Designations Compared (from 1 GHz upwards to 100 GHz)

Waveguide Bands	1.12	2.26	3.95	8.2	12.4	18.0	26.5	36	46	56	100
Designation Letter	L	S	C	X	Ku	K	Ka	Q	V	W	—
NATO Bands	1	2	3	4	6	8	10	20	40	60	100
Designation Letter	D	E	F	G	H	I	J	K	L	M	—

Also note that V-band is being defined in the usual way associated with millimeter-wave link applications, that is, not the relatively broad IEEE definition that extends from 40 to 75 GHz.

It is vital to be able to convert from frequency to wavelength (and vice versa). The standard general expression is

$$\lambda = \frac{v}{f} \text{ (m)} \qquad (1.1)$$

here λ is the wavelength (m), v is the velocity of the wave (m/s), and f is its frequency (Hz). For electromagnetic waves in free space or air, v is very close indeed to 3×10^8 m/s. This 3 factor accounts for the extensive use of 3 in the frequency bands cited in Table 1.2, because this makes for relatively rapid calculations of associated wavelengths.

At high frequencies, notably through superhigh frequency and extrahigh frequency, working in mm.GHz units is more practical and appropriate so that (1.1) becomes

$$\lambda = \frac{300}{f \text{ (GHz)}} \text{ (mm)} \qquad (1.2)$$

where f is substituted directly in gigahertz. A signal at 30 GHz, for example, has a free-space wavelength of 10 mm. Radio waves traveling in dielectrics have their wavelengths reduced by the square root of the dielectric's permittivity ε_r. So the wavelength in this case is given by the following modified version of (1.2), that is:

$$\lambda = \frac{300}{f \sqrt{\varepsilon_r}} \text{ (mm)} \qquad (1.3)$$

This means that if, for example, a 30-GHz signal is traveling entirely within a dielectric that has a permittivity of 2.3 (e.g., circuit board) the wavelength is reduced to 6.59 mm. Where RF design is required involving signals on circuit boards, ceramic substrates, or semi-insulating semiconductors, the general concept regarding wavelength calculation is vital. This book includes many examples of this situation.

For microstrip or any other quasi-TEM transmission, which is very important in microwave or millimeter-wave design, some of the electromagnetic fields extend into the air as well as the substrate. In these circumstances, (1.3) still applies, albeit with the relative permittivity replaced with a quantity called the effective permittivity, ε_{eff}. Equation (1.3) then becomes

$$\lambda = \frac{300}{f \sqrt{\varepsilon_{eff}}} \text{ (mm)} \qquad (1.4)$$

There is much more information on this important topic in Chapter 3.

1.5 Software-Defined Radio and Cognitive Radio

A detailed systems-level block of a software-defined radio (SDR) is shown in Figure 1.5.

The first (overall) observation to be made is the fact that most of the hardware is digital, including software processing. Only the block to the far left-side (flexible RF hardware) is RF. Most of the circuit functions are described elsewhere in this book (notably Chapters 5 and 11), but several other aspects need clarification:

- Flexible RF hardware refers to RF circuits that come under the control of the processing software.
- CORBA (Common Object Request Broker Architecture) is software that provides dynamic (strategic) decisions within a network.
- Virtual radio machine (alternatively, radio virtual machine) provides the SDR with portability and platform reconfigurability.

Most SDRs operate at frequencies below around 6 GHz and are therefore readily implemented mainly using silicon complementary metal-oxide semiconductor (CMOS) ICs. However, in some instances, Gallium arsenide (GaAs) or Gallium nitride (GaN) chip technology may be required at the final front-end to provide greater efficiency and/or higher RF output power. SDR technology is used in the high-volume example of cell-phones (i.e., mobile phones).

In the late 1990s the concept of SDR was taken a radical step further with the early development of cognitive radio. Mitola's dissertation from 2000 [6] is widely regarded as the birth of cognitive radio as such, with Bostian [7] providing much detailed information on this subject.

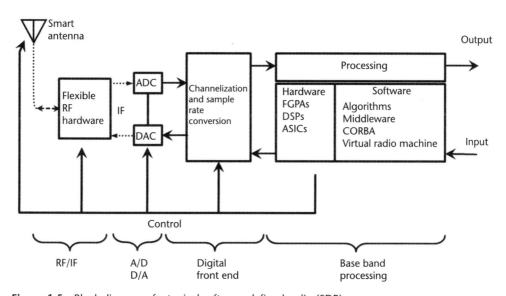

Figure 1.5 Block diagram of a typical software-defined radio (SDR).

One definition is that a cognitive radio (CR) is a radio system that has been programmed and dynamically configured to select the best RF channels locally available. A CR automatically detects the channels that are available within the RF spectrum and subsequently alters its internal parameters in order to accommodate the maximum possible number of live communications channels at the particular location. The process can be viewed as a type of dynamic spectrum management.

Originally CR was considered merely as an extension of software-defined radio, but it soon became evident that CR is much more than that.

Spectrum sensing for CR (notably across TV bands) is a very important research subject and major issues include the design of sufficiently high-quality devices and algorithms to enable the exchange of spectrum sensing data between nodes in the networks. It is abundantly clear that basic energy detectors are nowhere near sufficient for the task of accurately detecting the presence of signals at CR network nodes. Therefore more sophisticated spectrum sensing techniques are necessary. It is also clear that (as might be anticipated intuitively) increasing the number of cooperating sensing nodes progressively decreases the probability of false detection.

One approach that shows considerable promise is to implement a technique known as orthogonal frequency-division multiple access (OFDMA) so as to adaptively fill available RF bands.

1.6 The Challenge of Noise

In the broadest sense, noise comprises unwanted and apparently random perturbations that can potentially cause damage to audio or video content or cause errors in digital signals. Electrically, noise power covers a very wide spectrum and has several specific origins such as thermal, flicker ($1/f$) or phase noise. All these types of noise are important in characterizing and designing circuits and systems. Phase noise, in particular, is an important parameter in oscillators and many amplifier configurations.

Any receiver will have many specification characteristics and one of these is the minimum detectable signal (MDS). A concept of MDS is provided in the spectrum shown in Figure 1.6, which shows the MDS almost buried in noise.

Practically every component in an electronic circuit, active or passive, contributes to the noise and further details are provided in Chapters 2 and 3. Even basic resistors produce thermal noise and every transistor generates still more of this unwanted noise. Phase noise, especially close-in to the carrier, represents a very important quantity in this respect, notably in frequency sources but also in amplifiers designed for sensitive requirements. For a fundamental crystal-stabilized oscillator, the phase noise even as close as 100 Hz from the carrier (typically 100 or 140 MHz) can be as low as −144 dBc. However, the phase noise for most other types of frequency source is more like −70 dBc, even at frequencies much further removed from the carrier (e.g., 10 or even 100 kHz away).

Signal-to-noise ratio and also noise figure represent important specifications for amplifiers and receiver subsystems. Noise figure always increases with frequency and the first stage in any receiver is always the most sensitive. Alternatively, another quantity termed equivalent noise temperature can be used, particularly

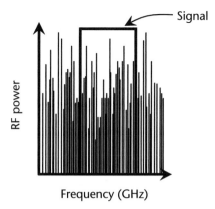

Figure 1.6 RF Signal almost buried in the noise.

in sensitive systems but also helpfully toward the analysis of noise in almost any system.

Much more detail is provided in Chapters 9 and 12.

1.7 RF Receivers

Until the twenty-first century, RF receivers were largely analog-based and took on the overall configuration shown in Figure 1.7.

Two bandpass filters are implemented in this receiver: the initial BPF and the image-reject filter. Oscillators are dealt with in Chapter 12 and mixers are covered in Chapter 13.

However, there are many systems that operate at RF and microwave frequencies up around 6 GHz and these types of systems increasingly demand extensive all-digital signal processing. The advancing frequency capabilities of analog-to-digital converters (ADCs) mean that after the LNA the next active circuit block an input signal to a receiver reaches is the ADC (ADCs are covered in detail in Chapter 11 and relatively briefly in Section 1.9).

The requirement for the initial BPF still very much exists, except that now the specification of that filter is much more stringent. In particular, a high degree of

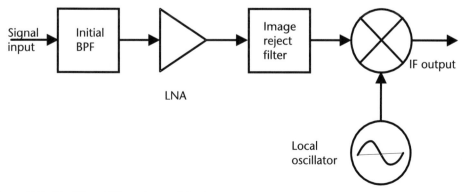

Figure 1.7 Traditional RF receiver architecture.

frequency agility is demanded (to cope with rapidly changing input channels) and a tuned BPF must now be designed. The initial sections of such a receiver chain are shown in Figure 1.8.

With the receiver architecture shown in Figure 1.8, the first two stages are similar to those shown in Figure 1.7, except (very significantly) the initial BPF must now be tunable. The second BPF is designed to substantially reduce any unwanted spurious components that may arise from the digital circuits. The ADC generates the digital bit streams required for digital signal processing. To the right-hand side of this system all signals are digital bit streams.

Software-defined radio (SDR) and cognitive radios, cited in Section 1.5, represent increasingly important examples of these classes of systems.

Channel multiplexers can readily be realized by inserting different BPFs into the arms of signal dividers, such as the Wilkinson power dividers described in Chapter 5. Each BPF is individually designed to pass the band of frequencies associated with a specific channel.

1.8 RF Filters

Frequency filters, notably bandpass filters (BPF) but also lowpass filters (LPF), are critical functional blocks in all RF systems. In particular, a BPF is almost always required between the antenna and the first low-noise amplifier (LNA) in a receiver.

Traditionally, this initial BPF has fixed parameters: center frequency, channel bandwidth, attenuation levels, and so on. This type of BPF will remain important, especially as operating frequencies are shifting upwards to include millimeter-wave bands.

The practical realization of any filter varies greatly, but many examples adopt planar technologies because these are consistent with both hybrid and MMIC/RFIC circuit approaches (see Section 1.2). Both lumped-element and transmission line technologies can be implemented and Figure 1.9 illustrates these options.

Figure 1.9(a) is representative of a lumped-element LPF. This layout could apply directly to a real (RF) LPF circuit, or it could be the low-frequency prototype that will ultimately lead to the design of the full RF LPF.

For general rule-of-thumb guidance where Figure 1.9(a) is actually an RF lumped-element LPF, the values of the components will define the frequency response:

- Through microwave frequency bands, the component values will be pF of capacitance and nH of inductance.

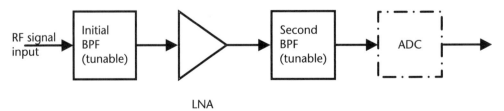

Figure 1.8 Digitally oriented RF receiver architecture.

1.8 RF Filters

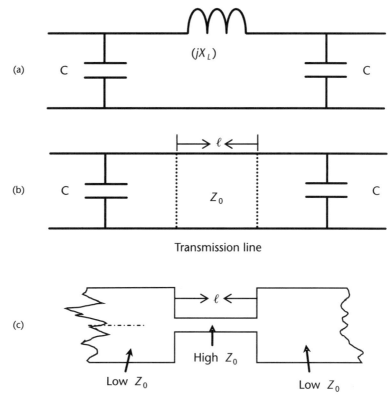

Figure 1.9 (a–c) Progression of a basic LPF lumped-component π-section through to a microstrip realization.

- Into the lower millimeter-wave bands, the component values will be more like fF of capacitance and maybe still some nH of inductance (possibly sub-nH).

However, the Q-factor (defined by the power losses) of the inductors in particular will usually limit the applicability of lumped components and instead lengths of transmission line are then adopted. The transmission lines are usually microstrip the circuit of Figure 1.9(a) first into Figure 1.9(b) (where the transmission line represents the inductor) and finally to Figure 1.9(c) where the inductor and the two capacitors are replaced by appropriately designed microstrip sections.

The microstrip transmission line approach is taken a stage further to design bandpass filters and the resulting structures typically take on the configuration shown in Figure 1.10.

Many practical hybrid and monolithic (MMIC) circuits can be designed implementing this type of BPF.

It is important to appreciate that all filters inherently suffer from the following defects in their spectral characteristics:

Figure 1.10 Typical bandpass filter realized using a cascade of half-wavelength coupled microstrip resonators (six-resonator BPF in an angled layout).

1. Finite insertion loss through the passbands (typically a fraction of a decibel);
2. Finite skirt insertion loss slopes (i.e., it is physically impossible to obtain sudden [infinitesimal] changes from passband to stopband).

1.9 ADCs and DACs

An analog-to-digital converter (ADC) accepts an analog (real-world) signal as input and this input is processed (i.e., converted) into a corresponding digital output signal. Symbolically an ADC is represented in any RF system as shown in Figure 1.11.

In many cases, the digital output comprises a 2's complement binary number that closely represents the analog input, although there are other possible digital representations. The sampling, quantization, and coding are especially critical electronic operations. There are many basic types of ADC, but only some of these are applicable to RF systems. Good examples of such ADCs include flash, folding, pipelined, and time-interleaved configurations. Details are provided in Chapter 11.

Following the digital signal processing it is necessary to convert the signal back to analog format and this function is performed by a digital-to-analog converter.

However, the raw output from a DAC is usually only a rough representation of the original analog signal, as indicated in Figure 1.12, and a lowpass reconstruction filter (LPF) is necessary after this output.

There are many specification points associated with DACs but for present purposes only one is considered: resolution.

A DAC's resolution is a strong function of the number of bits N involved in the digital input code. The resolution value expresses the number of different states possible for the DAC, and determines the minimum value of the step in voltage that the circuit can resolve.

More details on this together with several other specification points are presented in Chapter 11.

1.10 Oscillators, Mixers, and Frequency Converters

The broad scope considering radio frequency sources includes the following types of circuits and subsystems:

- Oscillators of various types, especially oscillators based upon quartz crystals;
- Dielectric resonator-based oscillators;
- Voltage-controlled oscillators (VCOs);

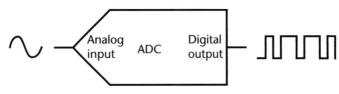

Figure 1.11 An ADC shown symbolically.

1.10 Oscillators, Mixers, and Frequency Converters

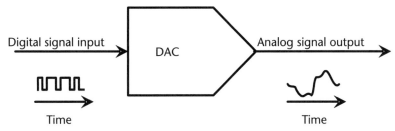

Figure 1.12 Basic schematic diagram of a DAC.

- Frequency multipliers;
- Frequency dividers;
- Frequency synthesizers (notably phase-locked-loop-based).

Many direct sources (also many high-stability master oscillators) comprise quartz-crystal-stabilized oscillators. Over recent years, various alternative technologies such as microelectronic mechanical systems (MEMS) have contended for application as frequency stabilizing elements, but concurrent developments in quartz crystals have predominated.

For many frequency sources, it is necessary to be able to electronically control the final output frequency and the basic concept is illustrated in Figure 1.13.

In practice, the tuned element is subject to extremely fast frequency variation by means of a varactor diode. The capacitance of such a diode alters as a function of the applied voltage and although this variation is highly nonlinear, linearization techniques can be applied. Varactor diodes are described in detail in Chapter 2.

Mixers are mainly required to deliver an output known as the intermediate frequency (IF) for further signal processing. There are many types of mixers (described in detail in Chapter 13), but the important example of the Gilbert cell is covered briefly here. The Gilbert cell mixer is also known as a four-quadrant multiplier because it mixes two signals by effectively multiplying them.

Silicon technology is most commonly encountered but silicon germanium (SiGe) bipolar complementary metal-oxide semiconductor (BiCMOS) and GaAs high electron mobility transistor (HEMT) approaches have also been demonstrated [5, 6].

This type of double-balanced mixer exhibits conversion gain and a superior noise performance compared with other mixer configurations (particularly when

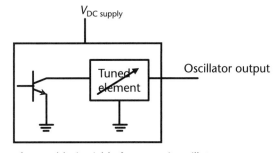

Figure 1.13 Concept of a tunable (variable-frequency) oscillator.

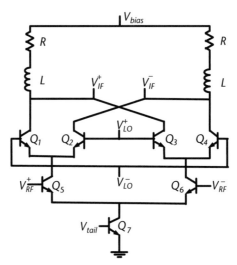

Figure 1.14 Basic structure of the Gilbert cell mixer, implemented using BJTs but omitting the required input/output baluns [8]. (© Artech House, 2016.)

SiGe heterojunction bipolar transistors [HBTs] are implemented). Being transistor-intensive, this circuit is also highly suited to MMIC/RFIC realization.

Extensive further details regarding mixers, including the Gilbert cell, are provided in Chapter 13. Occasionally, it is necessary to divide down a particular frequency and various options are available for this purpose. A good example of the requirement is within a phase-locked (PLL)-based frequency synthesizer, which is now briefly described here because these are of particular importance in communications systems.

The basic configuration is shown in Figure 1.15.

Key features of this synthesizer include:

1. The reference oscillator is usually a quartz crystal-based circuit providing very high stability and the lowest possible phase noise (see Section 12.3).
2. The phase detector will almost certainly comprise the balanced detector (two-diode) configuration, described in Chapter 13.
3. The amplifier can comprise an operational type as cited elsewhere in this book (e.g., in Chapter 11).
4. Lowpass (loop) filters are described in Chapter 6.
5. VCOs and frequency dividers are described in Sections 12.5 and 12.7, respectively.

The major key to the operation of this PLL is the phase detector where the phases of the reference oscillator and the divided output frequency are compared. Bearing this critical feature in mind, major aspects in the operation of this loop are described in Chapter 12.

The fractional-N subsystem represents a very important variation of this basic synthesizer and full details of this are also given in Chapter 12.

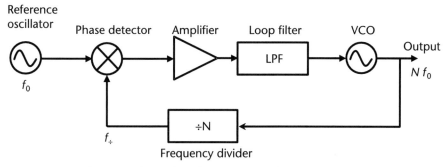

Figure 1.15 Basic subsystem of a PLL-based frequency synthesizer.

1.11 Semiconductor Device Requirements

Exactly what is meant by the term semiconductors? To an industrialist, semiconductors is a term embracing discrete devices and integrated circuits (diodes, transistors, and ICs), whereas to a professor of electronics, semiconductors mean the basic materials from which diodes, transistors, and ICs are manufactured. Either way, a basic understanding of the semiconductor materials and of the devices (the diodes, transistors, and ICs) are both essential because these underlie the detailed RF circuits and systems.

It is being assumed that the reader of this book is already well aware of diodes, transistors, and ICs as these apply within lower-frequency systems or within computers and digital control configurations. The purpose here is to focus on the special requirements regarding RF semiconductors.

1.12 Semiconductor Manufacturing

This is a highly specialized subject in its own right; therefore, only a brief description is provided here. A very good detailed coverage is supplied in [2, 9].

However, it is not particularly important for RF communications technologists to be familiar in great detail with semiconductor manufacturing.

The approach depends on the exact nature of the semiconductor products to be manufactured, although all processes have one thing in common: they all begin with a high-purity, single-crystal boule of the intrinsic semiconductor (i.e., silicon, gallium arsenide, gallium nitride, or other semiconductor material). In the case of silicon, this single-crystal boule is often around 15 cm in diameter. However, for GaAs or GaN, the boule diameters are generally much smaller, less than 10 cm. The issue of boule diameter is a serious one because this diameter determines the approximate number of die that can be made on the wafer (the top portion of the boule). In turn, this leads to the production yield of good die, which has immediate economic implications.

Photolithography, successive selective diffusion (of dopant materials), and selective metallization are all essential steps toward the manufacture of any RF semiconductor device. For MMICs (or RFICs), it is necessary to add the realization of vias connecting between various nodes on the chip.

The manufacturing of SiGe BiCMOS MMICs is particularly specialized because this requires many more mask stages than the other technologies. As a result, this specific technology is only economically viable for high-volume applications.

1.13 Diodes and Transistors

The diode is the most basic RF semiconductor device, and it remains of great importance in many circuits.

Schottky-barrier diodes (often just called Schottkys) embody a metal anode that directly joins a semiconductor cathode. The physical structure and commonly used circuit symbol are shown in Figure 1.16.

In Figure 1.16 an N-doped semiconductor is shown. This is the most common form because the mobility of electrons (hence, N) is always much higher than that of holes. This, along with many other aspects, is fully described in Chapter 2.

Various other types of RF diodes are available, notably PIN diodes and varactor diodes. These, including noise characteristics, are described in Chapter 2.

Basic aspects regarding bipolar junction transistors were described earlier (notably Figure 1.3 in which C refers to the device's collector, B refers to its base, and E refers to the emitter). An NPN transistor is selected because most RF transistors adopt this structure, dominated by the relatively high mobility N-doped semiconductor. These general types of devices are available as some particularly sophisticated structures, such as the GaAs HBT shown in Figure 1.17.

In contrast, the relatively straightforward field-effect transmitter (FET)-type transistors involved in CMOS transistors are also very important.

The basic CMOS logic inverter represents an important example of how two contrasting types of MOS transistor are interconnected to form a fundamental type of circuit configuration, shown in Figure 1.18.

It is important to observe the small circle symbol on the (gate) input to the upper transistor, which means that this p-type metal-oxide semiconductor (PMOS) transistor's gate directly connects to an N-well region. In contrast, the lower n-type metal-oxide semi-conductor (NMOS) transistor's gate directly connects to the P-type substrate. Further details are beyond the scope of this book. When both transistors have minimum feature dimensions down into the submicron levels (increasingly nanometers), these types of circuits can be designed to process low-power microwave and millimeter-wave signals.

RF CMOS and its derivatives now represent a mainstream RF technology that can be adopted for the relatively low-power portions of MMICs/RFICs. In order to increase the operating speed (hence also frequency), it is also possible to add one (sometimes two) bipolar transistors to a CMOS circuit stage.

Figure 1.16 Schottky-barrier diode: (a) schematic of structure, and (b) circuit symbol.

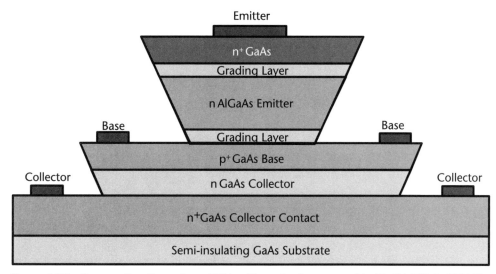

Figure 1.17 Cross section through an NPN gallium aluminum arsenide (GaAlAs)/GaAs HBT (the actual size has all dimensions in nanometers; the largest dimension can approach 100 nm).

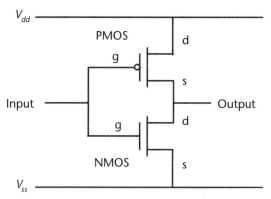

Figure 1.18 Basic CMOS logic inverter circuit.

GaAs HEMTs, GaAs pseudomophic HEMTs (pHEMTs), and GaN HEMTs are also vital transistor types that are designed into several types of microwave and millimeter-wave circuits.

1.14 Hybrid Circuits and MMICs

The basic requirement for any solid state amplifier is the internal transistor, and these fundamental semiconductor devices are described in Chapter 2 of this book. While the main thrust of technology choice is toward MMIC/RFIC realizations, discrete transistors are required where:

- There is a need for relatively high output power, generally upwards of several tens of watts (CW) or kW (pulsed);

- Scenarios where the RF output power may be relatively low (generally below a few tens of watts), but custom or low production rate designs are the order of the day.

As an example, a 100-W RF power amplifier (RFPA) operating around 2 GHz typically implements one or more discrete GaN HEMTs and such a design would almost certainly take the hybrid circuit route, most likely on a polymer-based circuit board with excellent heat-sinking.

In contrast, an RFPA required to provide a 10-W output at moderate frequencies would very likely be designed in MMIC format—provided the production rate is at least several thousand pieces. The transistor process in this case will depend mainly on the signal frequencies involved: typically GaN HEMTs for lower microwave frequencies although more likely GaAs pHEMTs for designs around or above 26 GHz.

However, the strong trend is toward silicon transistor processes for the lower-power scenarios. All these types of processes are described in Chapter 2.

As hinted above, the selection of the transistor process depends critically on the operating frequency. This feature is a consequence of the internal and parasitic reactive elements associated with every transistor, which again are described in Chapter 2. These reactive effects strongly influence the RF output power and power gain as functions of frequency for any transistor together with RFPAs implementing these devices. In general, both available output and power gain tend to decrease with increasing frequency.

An example of a MMIC-based RFPA is shown in Figure 1.19.

Lange couplers can be seen at both the input and output sides of this chip. The Lange coupler is described in detail in Chapter 4. Where the circuit is integrated (as in this example), all the bond connections within the Lange couplers are formed using air bridges.

1.15 The Challenge of RF Power Amplification

Every basic power amplifier will always exhibit signal distortion, whatever the technology. This distortion is a fundamental issue that has exercised RFPA designers

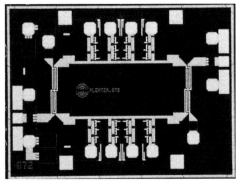

Figure 1.19 E-band MMIC implementing Lange couplers at both input and output ends. (Courtesy of Plextek RFI.)

over the years. The distortion arises mainly because of the inherent nonlinearity in the internal transistor current-voltage characteristics, which are particularly pronounced under large-signal (PA) conditions. An example of the resulting power amplifier input-output power transfer characteristic is shown in Figure 1.20.

It can be seen from Figure 1.20 that a somewhat more linear behavior is obtained by backing off the input so the output power is also automatically backed off, but clearly there is still a substantial amount of distortion.

Predistortion techniques are usually applied to compensate for this distortion, notably digital predistortion (DPD).

Another important parameter associated with power amplifiers is the overall efficiency and this requirement has yet another strong bearing on the choice of amplifier configuration. Details regarding efficiency, DPD, and further aspects of RFPAs are provided in Chapter 10.

1.16 Electronic Design Automation

Electronic design automation (EDA) is a vital, ongoing necessity for electronic circuit and system design and simulation. Several companies have embraced the special requirements of RF-EDA, and in this section brief summaries are provided of each vendor's main offerings. First, here is some general information regarding EDA software tools.

The main aim of any EDA package is to enable the full design of a circuit or subsystem and then to simulate its performance. Toward this aim, every EDA package requires the following minimum precisely detailed inputs:

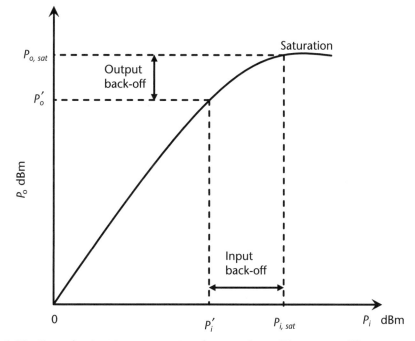

Figure 1.20 General output-input power transfer curve for an RF power amplifier.

- The LC overall final circuit or subsystem specifications (e.g., center frequency, bandwidth, noise figure, power levels);
- The basic technology to be used, especially whether hybrid of monolithic (i.e., MMIC/RFIC, immediately this will decide as to whether millimeter or micrometer dimensions are appropriate);
- Following on from the last input, details concerning the substrate, notably thickness and permittivity;
- The LC specifications regarding active devices, mainly transistors: discrete devices for hybrid circuits, processes where MMICs are concerned (e.g. CMOS, GaAs pHEMT, GaN HEMT).

In most instances, the software will already contain extensive libraries of data on various typical components and processes. Design routines will be available for transmission lines (coplanar waveguide or microstrip), for specific types of passive circuit structures (e.g., baluns, Lange couplers, or Wilkinson dividers) and also for lumped components such as capacitors, inductors and resistors.

All the required data must be inputted. Outputs will usually include the final circuit layout together with graphical plots of performance, typically to a base of frequency (i.e., simulation results).

References

[1] Khanna, A., "mmWaves Hit the Highway," *Microwave Journal*, August 2017, pp. 22–42.

[2] Kingsley, N., and J. R. Guerci, *Radar RF Circuit Design*, Norwood, MA: Artech House, 2016.

[3] Brookner, E., "Metamaterial Advances for Radar and Communications," *Microwave Journal*, November 2016, pp. 22–42.

[4] Ghosh, A., "The 5G mmWave Radio Revolution," *Microwave Journal*, September 2016, pp. 22–36.

[5] Kovitz, J. M., J. H. Choi, and Y. Rahmat-Samii, "Supporting Wide-Band Circular Polarization," *IEEE Microwave Magazine*, July/August 2017, pp. 91–104.

[6] Mitola, J., "Cognitive Radio: An Integrated Agent Architecture for Software Defined Radio," Ph. D. dissertation, Dept. Tech. Royal Inst. Tech., Sweden, 2000.

[7] Bostian, C. W., N. J. Kaminski, and A. S. Fayez, *Cognitive Radio Engineering*, Edison, NJ: SciTECH Publishing/IET, 2016.

[8] Camarchia, V., R. Quaglia and M. Pirola, *Electronics for Microwave Backhaul*, Norwood, MA: Artech House 2016.

[9] Edwards, T., and M. Steer, *Foundations for Microstrip Circuit Design*, 4th ed., New York: John Wiley & Sons, 2016.

CHAPTER 2

RF Semiconductors

2.1 Introduction

Exactly what is meant by the term semiconductors? To an industrialist, semiconductor is a term embracing discrete devices and integrated circuits (diodes, transistors, and integrated circuits [ICs]), whereas to a professor of electronics the word semiconductor means the basic materials from which diodes, transistors, and ICs are manufactured. Either way, a basic understanding of the semiconductor materials and the devices (the diodes, transistors, and ICs) is essential, and this is the purpose of this chapter.

It is being assumed the reader of this book is already well aware of diodes, transistors, and ICs as these apply within lower-frequency systems or within computers and digital control systems. Several texts covering fundamentals of semiconductors are readily available, Sze and Ng [1] and Shur [2], for example.

The purpose here is to focus on the special requirements regarding RF semiconductors. This chapter begins by studying key aspects of semiconductor materials and then majors on a detailed examination of diodes, transistors, and ICs for RF applications. Many useful aspects are also provided in [1–3].

2.2 Semiconductor Materials

It is very well known that the great majority of electronic systems are based around silicon. This fundamental semiconductor material forms the basis for almost all the ICs designed into systems ranging from computers through to iPads, mobile phones (cell phones), and a wide variety of other devices. However, although silicon is increasingly important in RF communications systems, it is certainly not the only semiconductor material required in this context.

To explore why this is the case, it is necessary to review some fundamental aspects and characteristics associated with an important range of semiconductor materials. As well as silicon, several other types of semiconductor materials are considered, notably gallium arsenide (GaAs), gallium nitride (GaN), indium phosphide (InP), and silicon germanium (SiGe) all of which are termed compared semiconductors.

There are many important aspects of semiconductors and the concepts of bandgap and drift velocity are particularly significant.

2.2.1 Bandgap

In any solid material, including semiconductors, if N valence electrons (all having the same energy) are combined to form bonds, then N possible energy levels will be the result. Exactly half of these energy levels will be decreased in energy while the remaining half will exhibit increased energy.

However, a statistical situation will exist whereby each half cannot simply contain exactly identical energy levels and instead in practice there is always a statistical distribution of electron energy level occupancies. For semiconductors, this distribution follows the Fermi-Dirac function, which is exponential and leads to the important exponential current-voltage relationships that characterize most diodes and transistors. Further details explaining this are beyond the scope of this text, but it is useful to appreciate the concept of a bandgap diagram. A simple, basic example is presented in Figure 2.1.

In Figure 2.1 E_F is the Fermi level (middle of the bandgap = $E_c - E_v$).

It requires a specific amount of energy to cause an electron to become released from the valence band, cross the bandgap, and hence be available for conduction. The energy difference between the top of the valence band and the bottom of the conduction band is called the bandgap and bandgaps for a range of important semiconductor materials are quoted in Table 2.1.

Bandgaps (also mobility, considered later) are amongst the important criteria leading to the choice of semiconductor material that may be used as the basis for any semiconductor component. Also, very importantly, additional materials can be introduced (such as aluminium) so as to alter the bandgap according to requirements. This is known as bandgap engineering.

Semiconductor materials are generally classed as being either:

- Narrow bandgap semiconductors: all except GaN are within this category (see Table 2.1);

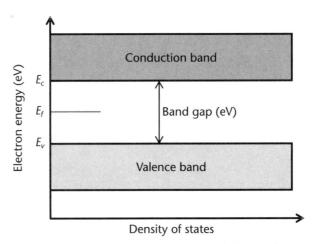

Figure 2.1 Basic concept of the energy band structure in a solid material.

Table 2.1 Energy Bandgaps for Various Selected Semiconductor Materials

Semiconductor Material	Energy Bandgap (eV) at Room Temperature ~300K
GaAs	1.42*
GaN	3.36**
InP	1.35
Silicon (crystalline) (Si)	1.12
Silicon-Germanium (SiGe)	0.67 to 1.11***

Notes: *1.5 eV at a field strength of 4 kV/cm then down to 1.0 eV at higher field strengths; **Decreases to 2.9 eV at a field strength of 150 kV/cm and reduces further to 1.5 eV at higher field strengths; ***The extremes of SiGe can either be very like germanium or much more like silicon, dependent on the alloy choice, hence the extremes of possible bandgaps.

- Wide bandgap (WBG) semiconductors. GaN is the prime example, although there are some other WBG materials.

Regarding field strength units of kV/cm, bear in mind a distance of, say, 0.5 mm may be the case with practical semiconductor devices, and 10 kV/cm translates to 1 kV per mm (i.e., 500V across a 0.5-mm distance).

Semiconductor devices, particularly transistors, comprising narrow bandgap semiconductor devices imply relatively low DC supply voltages for circuits composed of these, whereas wide bandgap (WBG) semiconductors tend to mean transistors embodying these require relatively high DC supply voltages. Most transistors based on GaN semiconductor require DC supply voltages of the order of some tens of volts, whereas many circuits using GaAs or silicon transistors will operate using DC supply voltages as low as 3V.

2.2.2 Drift Velocity

In any semiconductor material the electrons or holes will accelerate under the attraction of an applied electric field. The speed of the electrons is termed the drift velocity and the ratio of this drift velocity to the applied electric field is termed the mobility. This mobility, symbol μ, is a very important parameter for selecting semiconductors that could be candidates for RF applications. The term drift velocity is somewhat misleading in that it may suggest the electrons travel slowly and even randomly, whereas they travel at high speed.

For real devices, various parts of the diodes or transistors are doped N-type (electron-intensive) or P-type (hole-intensive) by the introduction of various doping materials into the intrinsic (high-purity) semiconductor.

For reasons that are beyond the scope of this book the mobility of electrons μ_e is always much higher than that of holes μ_h, which means electrons will travel much faster than holes under the same electric field. This is the reason why most RF or microwave diodes and transistors are designed so that the critical electron-transport sections are N-type rather than P-type. Electron mobilities for a range of important semiconductor materials are quoted in Table 2.2.

Table 2.2 Electron Mobilities for Various Selected N-type Doped Semiconductor Materials

Semiconductor Material	Approximate Electron Mobility μ_e ($cm^2V^{-1}s^{-1}$) at Room Temperature
GaAs	8,000*
GaN	990 to 2,000**
InP	5,400 (max.)
Silicon (Si)	1,400
Silicon germanium (SiGe)	1,800

Notes: *Can be as low as 2,500 $cm^2V^{-1}s^{-1}$ at high electron concentrations. **Can be as high as 10,000 $cm^2V^{-1}s^{-1}$ at moderate electron concentrations.

From the data in Table 2.2, it is clear that there is no outright winner in terms of mobility. InP is rarely used in commercial RF semiconductor devices on account of important issues such as dielectric loss and wafer cost, which prohibit its use except in the most demanding applications.

GaAs benefits from a notably high electron mobility and over several decades this semiconductor material has been the top choice with RF and microwave engineers. For many years since the 1970s, practically anything microwave meant the use of GaAs for most diodes, transistors and integrated circuit chips. GaAs devices such as GaAs pHEMT transistors (and ICs implementing these) remain important for practical twenty-first-century RF circuits. Further details are provided later here.

2.2.3 Resistors Made from Semiconductors

ICs designed for digital applications (processors, gate arrays, memories) are generally transistor-intensive and do not require any passive components such as capacitors, inductors, transmission lines, or resistors. However, ICs designed for RF/microwave applications are radically different and it is vital for such passive components to be designed. Chapter 3 deals with these requirements in some detail but at this point, armed with a knowledge concerning mobilities, in particular, it is appropriate to examine the design of a semiconductor-based resistor.

The first question to pose and to answer is how to calculate the conductivity of a semiconductor material. The following equation provides this:

$$\sigma_e = n_e q \mu_e \quad (2.1)$$

in which σ_e, n_e, and μ_e are, respectively, the conductivity, electron concentration, and mobility of the N-type semiconductor material and q is the charge on an electron (1.601×10^{-19} C).

Finally, the resistance R of the rectangular strip is calculated using the classic formula:

$$R = \ell / \sigma_e A \quad (2.2)$$

where ℓ is the length of the strip and A is its cross sectional area.

2.2 Semiconductor Materials

As an example, calculate the length required of a rectangular strip resistor, value 200Ω, to be integrated within a SiGe-based RF IC. There are 10^{15} electrons per cubic centimeter in the conduction band and the resistor measures 300 μm × 150 μm in the cross section.

For the solution, first calculate σ_e using (2.1) and with mobility from Table 2.2:

$$\sigma_e = 10^{15} \times 1.601 \times 10^{-19} \times 1.8 \times 10^3$$

where μ_e for SiGe was obtained using Table 2.1. It must be checked that all the units are consistent, leading to the final units: S/cm.

This calculates to $\sigma_e = 0.288$ S/cm.

Next rearrange (2.2) to express the length of the resistor strip ℓ and substitute all the quantities, giving:

$$\ell = \sigma_e A R$$
$$\ell = 0.288 \times 3 \times 10^{-2} \times 1.5 \times 10^{-2} \times 200 \ \{units:cm\}$$

hence, $\ell = 2.594 \times 10^{-2}$ cm (i.e., $\ell = 259.4$ μm).

The length of this resistor on the SiGe substrate is ~259 μm.

2.2.4 Electron Speed and Transit Time

Typical electron velocities within a transistor are in the order of 10^5 m per second. The typical linear dimensions over which the electronics will transit are in the order of a 1 μm.

It is therefore very straightforward to determine the order of magnitude of the transit time of electrons traveling within the active region of a transistor. This transit time τ is:

$$\tau = {1 \times 10^{-6}}/{10^5} \ \{units: seconds\}$$

$\therefore = \tau = 10^{-11}$ seconds – or 10 ps

This is a fairly typical time interval, applying to microwave electronics.

It is however vital to appreciate that other parameters associated with transistors and related types of devices restrict the behavior and limit the device's time-domain characteristics and frequency response.

2.2.5 Some Further Important Properties of Semiconductors

In a later section of this chapter monolithic microwave integrated circuits (MMICs) or radio frequency integrated circuits (RFICs) are covered in some detail. For these technologies, the substrates are the semiconductor wafer materials and therefore relevant material properties beyond those described above are of importance. Table 2.3 provides some data of this nature.

All the parameters quoted in the above tables are important in understanding semiconductor devices. In the case of Table 2.3, the properties listed are vital input

Table 2.3 Properties of Various Selected Semiconductor Materials Relevant to Their Use as IC Substrates

Semiconductor Substrate	Relative Permittivity (ε_r)	Dielectric Loss Tangent (tan δ at 10 GHz)	Approximate Thickness h(mm)	Surface State Roughness (μm)	Dielectric Strength (kV/cm)	Thermal Conductivity (W cm^{-1}K^{-1})
GaAs	12.85	6	0.5	0.025	350	30
GaN	8.9	—	0.5	0.025	4,000	140
InP	12.4	5	0.6	0.025	350	40
Si (high resistivity silion [resistivity >2 kΩ.cm])	11.9	≈1	0.36	<0.001	300	120
SiGe	≈13	—	—	—	220	84

values for the design of passive components and interconnections on semiconductor substrates (MMICs, RFICs). Details are provided in Chapter 3.

2.2.6 Semiconductor Manufacturing

This is a highly specialized subject in its own right; therefore, only a brief description is provided here. A very good detailed coverage is available in [4].

However, it is not particularly important for RF communications technologists to be familiar in detail with semiconductor manufacturing. The approach depends on the exact nature of the semiconductor products to be manufactured, although all processes have one thing in common: they all begin with a high-purity, single-crystal boule of the intrinsic semiconductor (i.e., silicon, gallium arsenide, gallium nitride, or other semiconductor material). In the case of silicon, this single-crystal boule is often around 15 cm in diameter. However, for GaAs or GaN, the boule diameters are generally much smaller, <10 cm. The issue of boule diameter is a serious one because this diameter determines the approximate number of die that can be made on the wafer (the top portion of the boule). In turn, this leads to the production yield of good die, which has immediate economic implications.

Photolithography, successive selective diffusion (of dopant materials), and selective metallization are all essential steps toward the manufacture of any RF semiconductor device. For MMICs (or RFICs), add the realization of vias connecting between various nodes on the chip.

The manufacturing of SiGe BiCMOS MMICs is particularly specialized because this requires many more mask stages than the other technologies. As a result, this specific technology is only economically viable for high-volume applications. SiGe BiCMOS is discussed in some detail in Section 2.4.4.

2.3 Semiconductor Diodes (RF-Oriented)

2.3.1 Some Semiconductor Junction Diode Fundamentals

Any two-terminal semiconductor junction device is termed a semiconductor diode. The simple basic structure and its common circuit symbol are shown in Figure 2.2.

2.3 Semiconductor Diodes (RF-Oriented)

Figure 2.2 Semiconductor junction diode: (a) schematic of structure and (b) circuit symbol.

When a positive DC voltage is applied to the anode (left side in Figure 2.2[a]), a forward current I_F flows from P to N (follow the anode arrow in Figure 2.2[b]). This current varies exponentially as the voltage across the junction increases, as indicated in (2.3):

$$I_F = I_{SAT}\left(e^{\alpha V} - 1\right) \quad (2.3)$$

where I_{SAT} is the reverse-bias saturation current, V is the (DC) forward voltage, and the coefficient α is expressed as follows:

$$\alpha = q/{mkT} \quad (2.4)$$

in which q is the charge on an electron (1.601×10^{-19} C), k is Boltzmann's constant (1.38×10^{-23} J.K^{-1}), T is the absolute temperature (K), and m, the ideality factor, is a number between 1 and 2 dependent on the structure of the diode (notably the semiconductor material used).

Equation (2.3) is often referred to as the Shockley Ideal Diode Law.

The general semiconductor diode I/V characteristic is shown in Figure 2.3.

There are two important features in this overall characteristic:

1. The first quadrant, which is generally known as the forward-biased characteristic;
2. The third quadrant, which is generally known as the reverse-biased portion of the overall characteristic.

Depending on the detailed structure of the diode, for most RF diodes, the forward-biased region is characterized by DC current levels increasing through mA

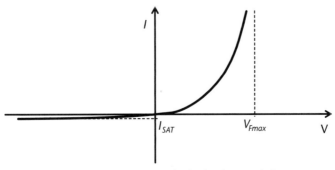

Figure 2.3 The general semiconductor junction diode I/V characteristic.

values, while corresponding DC voltages increase through hundreds of mV values, up the maximum voltage $V_{F,\max}$ of around 0.7V or 0.8V.

Contrastingly, for most RF diodes the reverse-biased region is characterized by the DC current level saturating at a maximum value I_{SAT} in the order of μA or even nA. Across this region DC voltages can easily reach several tens of volts before voltage breakdown occurs. This voltage breakdown value is often used in diodes specially designed to deliberately limit DC voltages (Zener diodes). However, this area is outside the scope of this book.

Schottky-barrier diodes (often just called Schottkys) embody a metal anode that directly joins a semiconductor cathode. The physical structure and commonly used circuit symbol are shown in Figure 2.4.

In Figure 1.16 an N-doped semiconductor is shown. This is the most common form because the mobility of electrons (N) is always much higher than that of holes as discussed earlier in this chapter. The shape of the general Schottky diode I/V characteristic is very similar to that of the junction diode, as shown in Figure 2.3. However, the value of $V_{F,\max}$ tends to be somewhat lower, around 0.6V to 0.7V.

The energy band characteristics of a Schottky diode can be developed from Figure 2.1, resulting in the diagram of Figure 2.4. In Figure 2.4, the Schottky barrier height, ϕ_B, is the difference between the interfacial conduction band edge E_c and the Fermi level E_F. It is notable that E_c decreases within the semiconductor, resulting in an asymmetric energy level situation. The dimension z is the distance progression through the structure from the metal anode to the cathode at the opposite end of the semiconductor.

2.3.2 P-I-N Diodes

By introducing an intrinsic layer of (most usually) silicon between the P and N materials of a junction diode, a switching (or sometimes attenuating) device is obtained. The intrinsic silicon has to be very slightly N-type or P-type because 100% intrinsic semiconductors are not realizable in practice.

PIN diodes are implemented in medium-to-high power systems, which make these devices of substantial importance. However, in lower-power communications systems, PIN diodes are only occasionally used. The great majority of communications systems operate using relatively low-power signal levels.

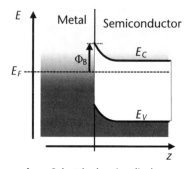

Figure 2.4 Energy-band diagram for a Schottky-barrier diode.

2.3 Semiconductor Diodes (RF-Oriented)

2.3.3 Varactor Diodes

On a DC I/V basis, the reverse-bias characteristic of a P-N junction is unremarkable, with simply the shape indicated in the third quadrant of Figure 2.3.

However, this situation is far from the end of the story regarding P-N junction diodes because under reverse bias the junction exhibits capacitance, and this capacitance C_T varies strongly with the DC applied voltage. The basic and general varactor capacitance equation is:

$$C_T = \frac{k_D}{(V_K + V_R)^q} \tag{2.5}$$

where k_D is a constant dependent on the diode structure; V_K is the knee voltage for the diode; V_R is the magnitude of the applied reverse voltage across the diode; and q is a parameter dependent on the type of junction. $q = 0.5$ for alloy junctions forming these types of diodes or 0.333 for diffused junctions.

This type of semiconductor diode is variously known as a varactor diode, or a varicap or a voltage-variable capacitance diode. The most commonly encountered term in the RF or microwave context is varactor diode and this terminology will be used consistently here.

From (2.5), it can be seen that the junction capacitance varies in a highly nonlinear fashion as a function of the applied reverse voltage and one example of the type of resulting curve is shown in Figure 2.5. This result is unfortunate from the viewpoint of the major application of varactor diodes, namely, voltage-controlled oscillators (Chapter 12) because a linear $C(V)$ behavior would have been ideal. Many techniques have been adopted in an effort to linearize the final $C(V)$ function.

A cross section through a typical varactor diode is shown in Figure 2.6.

To provide some idea regarding the behavior of this type of diode, consider the following sequence of events:

- As the reverse bias voltage is decreased, so the depletion layer narrows. This reduces the thickness of the depletion layer, which, in turn, increases the capacitance.
- As the reverse bias voltage is increased, so the depletion layer widens and the capacitance decreases.

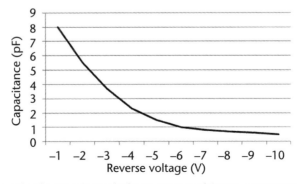

Figure 2.5 An example of a capacitance/voltage curve applying to a varactor diode.

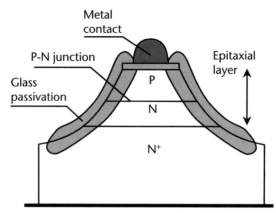

Figure 2.6 Cross section through a varactor diode. This would typically be of the order of a few millimeters in diameter and some hundreds of micrometers high.

- This behavior is highly nonlinear as indicated by (2.5).

The precise nature of the variation is determined by the doping densities and the size and geometry of the diode construction.

Most varactor diodes embody a doping profile that is highly abrupt (i.e., the P-N junction is extremely thin in comparison with the remaining dimensions). On each side of the junction, the doping concentration is maintained as constant as practically possible and these types of varactors are simply known as abrupt varactors. Due to the abrupt junction, the C-V characteristic follows an inverse square law.

There are also hyperabrupt varactors for which an inverse square law C-V characteristic also applies, although only over a portion of the C-V curve. An important consequence of this is that over a narrow range there is a linear frequency variation when the varactor is used in a VCO. However, the lowered Q-factor means that hyperabrupt varactor diodes can only be used up to fairly low microwave frequencies.

Circuit symbols for a varactor diode can either be displayed as Figure 2.7(a) or more simply (although much less meaningfully) by Figure 2.7(b).

A small-signal equivalent circuit for a varactor diode is shown in Figure 2.8.

In this equivalent circuit, R_s is a series resistance representing losses in the electrodes and R_V represents the losses associated with the depletion layer.

Varactor diodes are most often fabricated in silicon but sometimes in GaAs, which generally leads to higher frequency operation. Q-factors range from 1,200 to (exceptionally) 8,000, although these are usually measured at 50 MHz and will decrease (i.e., deteriorate) with increasing frequency.

2.3.4 Noise in RF Diodes

In common with all electronic devices, all types of RF diode generate noise as well as processing the desired signal. For example, both resistances indicated in Figure 2.7 will generate thermal noise, the mean-squared current being given by:

(a)   (b)

Figure 2.7 Circuit symbol options for a varactor diode.

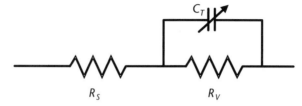

Figure 2.8 A small-signal equivalent circuit for a varactor diode.

$$\overline{i_R^2} = 4kTB/R \tag{2.6}$$

where all the parameters have been defined previously.

Also, the shot noise and the flicker noise generated by the perturbations in the P-N junction diode (also the Schottky diode) are expressed by:

$$\overline{i_d^2} = 2qBI_d + K_f \frac{I_d}{f} \tag{2.7}$$

where most of the parameters have been defined previously; also I_d is the DC current and f is the spot frequency. K_f is strictly a frequency dependent whose value tends to stay between 0.6 and 1.0.

Extra noise is delivered by varactor diodes (notably random variations in the capacitance) and this can seriously affect the phase-noise in VCOs. This is covered extensively in Chapter 12.

2.4 Transistors

2.4.1 Introductory Remarks

It is well known that the transistor was invented in 1947 by Shockley, Bardeen, and Brattain at Bell Labs. That landmark invention gave rise to an explosion in transistor development, leading to today's technologies, without which most of the world could barely function.

Many would understandably ask: What is a transistor? The answer to that basic question leads to a truly remarkable story. Until the invention of the integrated circuit (Kilby, Noyce) all transistors were discrete devices, and many twenty-first-century examples are indeed discrete transistors.

Currently, including RF electronics, the scenario includes discrete transistors as well as transistors embedded as parts of comprehensive circuits known variously as MMICs or RFICs. This chapter focuses on both discrete transistors and MMICs.

In general, it is important to appreciate that discrete transistors vary greatly in physical size, in power-handling capability and in maximum frequency of operation. The physical dimensions of high-power (kilowatt) devices may amount to many centimeters. However, transistors implemented in integrated circuits are naturally extremely small, having micrometer or even nanometer dimensions. Some transistors, designed for lower microwave frequencies such as L-band, can deliver RF power into the kilowatt range. In contrast, most RF transistors are designed to operate up to higher frequencies, increasingly towards even the terahertz bands. Apart from RF power-handling capabilities, low noise performance is also often very important.

2.4.2 High Frequency Circuit Models for Transistors

For design purposes, it is essential that equivalent models are available, representing transistor behavior in detail. First, it is vital to distinguish between small-signal and large-signal operation and subsequent modeling options:

- Small-signal modeling applies where the signal voltage and current excursions are much smaller than the applied DC values so that linear operation is maintained;
- Large-signal modeling is required where the above conditions for small-signal operation are inapplicable.

Given the significance of small-signal (linear) operation of most transistors and associated MMICs, it is understandable that small-signal models are more frequently encountered than large-signal models. In any instance, the model is required to closely represent the behavior of the transistor under the specified conditions including DC bias, frequency range, operating temperature, and so on.

There are two distinct approaches available in this important area:

- *Physics-based equivalent circuits:* Developed from fundamental transistor behavior (i.e., every component in the model is traceable to various electrical aspects of the structure);
- Black-box-based equivalent parameter sets that adequately (quantitatively) describe the transistor.

Some idea of physics-based equivalent circuits can be gained from Figure 2.10.
The equivalent circuit will comprise resistors, capacitors, and (for accurate representation at very high frequencies) inductance. In Figure 2.9, the circle with a central arrow represents a current generator and this is consistently required to account for any transistor's internal gain. Actual details are provided for various devices in the following sections.

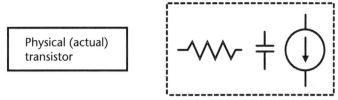

Figure 2.10 How the actual physical transistor can be represented by a small-signal equivalent circuit.

For black-box-based equivalent parameter sets, it is initially assumed that the transistor can be adequately defined in terms of various groups of two-port black-box parameters (Figure 2.11).

There are several sets of parameters that can be used to characterize a two-port black-box. Examples include the h- (hybrid) parameters, z-parameters, y-parameters, ABCD (or chain) parameters, and S-parameters. Y-parameters are occasionally used for characterizing RF transistors, but it is the S-parameters that are most favored under microwave conditions. The reason for this choice is that it is only the S-parameters that demand matched 50Ω terminations for their calculation and measurement. The reasons for the importance of 50Ω are described in Chapter 3 and details concerning these S-parameters are provided in Appendix B.

Relationships exist interrelating all the sets of parameters, and each set can be computed knowing the physics-based equivalent circuit elements.

There is one important exception to this situation, namely, X-parameters. These are special parameters that apply to the modeling of nonlinear effects observed when transistors operate under relatively large-signal conditions. The impact of transistor nonlinearity on RF power amplifier design and performance is fully discussed in Chapter 10. Further details regarding X-parameters are made available in Appendix B.

2.4.3 CMOS and Related Transistor Technologies

It is assumed that the reader will be well aware from lower-frequency and digital electronics that CMOS is exclusively silicon-based. In the twentieth century, it was almost unheard of to even remotely consider CMOS for RF applications because the highest speeds possible precluded such considerations. However, as MOS transistor feature sizes continued to shrink (and therefore operating speeds accelerated) so the possibilities of at least low-power microwave applications began to look feasible. Nowadays RF CMOS is well established for many microwave applications, especially where very densely packed mixed-signal RFICs are concerned. The basic CMOS logic inverter represents an important example of how two contrasting types of MOS transistor are interconnected to form a fundamental type of circuit configuration, shown in Figure 1.18.

Figure 2.11 Transistor considered as a black-box.

It is important to observe the small circle symbol on the (gate) input to the upper transistor, which means this PMOS transistor's gate directly connects to an N-well region. In contrast, the lower (NMOS) transistor's gate directly connects to the P-type substrate. Further details are beyond the scope of this book, but are available in references such as [1, 2]. When both transistors have minimum feature dimensions down into the submicron levels (increasingly nanometers), these types of circuits can be designed to process low-power microwave and millimeter-wave signals.

RF CMOS and its derivatives now represent a mainstream RF technology that can be adopted for the relatively low-power portions of MMICs/RFICs. In order to increase the operating speed (hence also frequency), it is also possible to add one (sometimes two) bipolar transistors to a CMOS circuit stage. Where such a BJT is NPN type then the base is often doped with germanium (Ge), which serves to shrink the bandgap substantially. This silicon-germanium alloy is named SiGe and further notes on this semiconductor are available in Section 2.2. The resulting overall technology is SiGe BiCMOS.

Compared to silicon alone, the addition of Ge:

1. Leads to a much higher value of the current gain ($\beta = I_c/I_B$);
2. Also leads to a much larger transition frequency (f_T).

SiGe BiCMOS technology is extensively and increasingly being adopted for RFICs going into high-volume applications such as automotive radars.

Some 130-nm SiGe BiCMOS transistors operate close to 1 THz, but a serious disadvantage is the very low breakdown voltage of typically around 1.7V. Adding BJTs to CMOS requires high-level multimasking fabrication which is relatively expensive and is certainly not cost-effective for low to moderate production runs. A study of MMICs and RFICs forms the final section to this chapter.

Laterally diffused metal-oxide-silicon (LDMOS) transistors are used for many ultrahigh frequency, high-power amplifier applications.

2.4.4 GaAs and GaN Field-Effect Transistors

In Section 2.2.2, it was pointed out that GaAs-based devices have been the mainstay transistors for RF engineering over many decades. Among such transistors, the GaAs metal-electrode-semiconductor field effect transistor (MESFET) was particularly important. In Section 2.3.1, the Schottky barrier is explained, and this applies to the metal-electrode-semiconductor input section of a GaAs MESFET.

Before considering any further specific technologies, it is useful to first identify the general equivalent circuit for a GaAs FET and then to analyze this so as to determine the transition frequency (along the same lines as for BJTs). The general small-signal equivalent circuit for a GaAsFET is shown in Figure 2.12 (common-source configuration).

The analysis follows that given above for a BJT, except all specific references are to the GaAs FET.

By definition:

2.4 Transistors

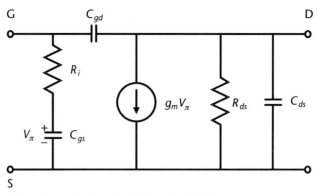

Figure 2.12 Small-signal π equivalent circuit for a GaAs FET.

$$G^{SC}_i = \left|\frac{i_D}{i_G}\right| = \left|\frac{g_m V_\pi}{i_G}\right| \tag{2.8}$$

Now,
$i_G/V\pi$ is the susceptance of the capacitance $B_{gs} = 2\pi f C_{gs}$
Substituting this into (2.8), setting $G^{SC}_i = 1$, and solving for frequency f_T yield:

$$f_T = \frac{g_m}{2\pi C_{gs}} \tag{2.9}$$

Example
A GaAs FET has transconductance (g_m) = 50 mS and gate-source capacitance (C_{gs}) = 0.2 pF. What is the (upper limit) transition frequency for this device?
Substituting directly into (2.9), it is found that f_T = 39.8 GHz.
This is almost double the f_T for a typical BJT. In spite of the much lower transconductance, it is the greatly reduced capacitance that dominates so that operation to at least around 30 GHz is possible.
While the GaAs FET remains a significant type of RF transistor, it has largely given way to two other distinctly different technologies: the GaAs HEMT (especially the pHEMT) transistor, and the GaN HEMT transistor family.

2.4.5 The GaAs HEMT and pHEMT

The acronym HEMT stands for high electron mobility transistor. The importance of mobility is explained and expanded upon in Section 2.2.2. An extensive treatment of HEMT is well beyond the scope of this book but some concept may be gained by considering the following brief description.
In a HEMT, the channel is effectively a quantum well. A quantum well is a depletion region within an undoped GaAs section that has been created by a heterogenous junction made using highly doped semiconductor materials. For example, this heterogenous junction could be P-N where both the P and N regions are highly

conductive. The channel is referred to as a two-dimensional electron gas (referred to as 2-DEG) and it has a very high conductivity.

The cross sectional structure of a GaAs HEMT is shown in Figure 2.13.

In this example, the n+ GaAs regions are highly conductive and the 2-DEG region occurs in the channel beneath the gate.

These effects can be greatly enhanced by ensuring the doped and intrinsic layers possess slight lattice mismatches. This situation is defined as pseudomorphic and it gives rise to the term pHEMT, hence a GaAs pseudomorphic HEMT or GaAs pHEMT.

The typical DC I_D/V_D characteristics for any FET-type transistor are shown in Figure 2.14.

It is clear from Figure 2.14 that the control parameter is V_g. Taking any constant drain voltage, increasing V_g causes the drain current I_D to increase up to a maximum value I_{max}. The fact that the characteristics only remain approximately linear over a range starting around $2V_{knee}$ is particularly significant and is followed up in Chapter 10.

In small-signal FETs, the voltages are in the order of volts while the drain currents are in the mA regime.

The noise generated in a FET can be significantly decreased by the alloying-in of a halogen (e.g., fluorine, chlorine, bromine, iodine, or astatine) into the gate dielectric [6].

2.4.6 The GaN HEMT

From Section 2.2, it should be clear that GaN is a truly revolutionary semiconductor material, notably for its application in RF electronics technology. Transistors based upon GaN likewise provide exceptional characteristics, and the main such transistor structure, shown in Figure 2.15, is the GaN HEMT.

In most designs, the substrate comprises silicon carbide (SiC). This is the preferred substrate material since it provides an excellent heat sink which is important for power devices. It is also possible to replace the n-GaN layer by n-AlN (aluminium nitride), but this will not be elaborated upon further here.

GaN HEMT transistors tend to be biased using relatively high-voltage supplies, such as 40V or more. They are mainly oriented toward medium-to-high power requirements and, as discrete devices with efficient heat sinks, can deliver in the order of many hundreds of watts continuous-wave (CW).

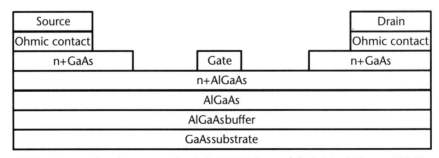

Figure 2.13 Cross sectional structure of a GaAs HEMT. (From: [5]. © Artech House, 2016.)

2.4 Transistors

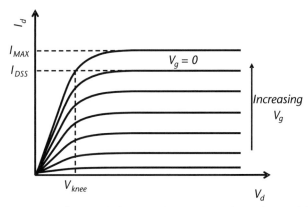

Figure 2.14 Output or DC characteristics on a FET-type transistor. (From: [5] © Artech House, 2016.)

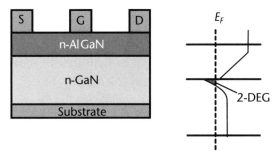

Figure 2.15 Cross sectional structure and energy-level diagram for a GaN HEMT.

In most instances, a large-signal equivalent circuit is appropriate and an example is shown in Figure 2.16.

Many GaN HEMTs are designed for use well into millimeter-wave, particularly small-signal devices. Most of these types of transistors form the major process in MMICs of various types. GaN HEMT developments in relation to distributed amplifiers are considered in detail by Ghavidel et al. [7].

2.4.7 Bipolar RF Transistors

While the device invented in 1947 was a germanium point-contact transistor, it soon became apparent that silicon bipolar junction transistors would be a much better option on account of performance capabilities, manufacturing operations and (crucially) final unit price to the customer. As a result of this, silicon bipolar junction transistors (BJTs) became the de facto standard for the great majority of transistors from the 1960s through to the end of the twentieth century.

Because BJTs remain important today for some significant RF circuits and systems, this technology is reviewed in this section. First, the basic structure of an NPN BJT is considered and the schematic for this (together with the circuit symbol) is shown in Figure 1.3.

In Figure 1.3, C refers to the device's collector, B refers to its base, and E refers to the emitter.

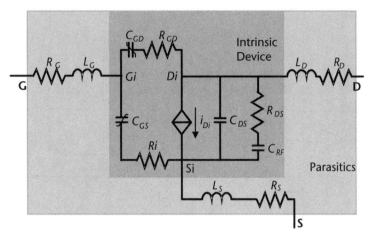

Figure 2.16 Large-signal equivalent circuit of a GaN HEMT.

An NPN transistor is selected because most RF transistors adopt this structure, dominated by the relatively high mobility N-doped semiconductor.

For any transistor a family of curves can be plotted indicating, on a DC basis, output current versus output voltage where an input parameter (current or voltage) is the control parameter. In the case of a BJT, this family of curves comprises collector current as the output current, collector-to-emitter voltage V_{ce} as the output voltage, and base current I_b as the control parameter.

In a practical high-power RF BJT, I_c could amount to several amps, V_{ce} might be several tens of volts, and I_b may well amount to some tens or hundreds of mA. For lower-power devices these values would be more like several mA, a few volts, and moderate values of I_b, typically some tens or hundreds of μA.

Transistor equivalent circuits are introduced in Section 2.4.2 (Figure 2.9) and the general small-signal hybrid-π equivalent circuit for a BJT is shown in Figure 2.17 (common-emitter configuration).

This circuit is referred to as hybrid-π because the series resistance disrupts what would otherwise be a pure-π configuration. The voltage V_π is shown with polarization to account for the current generator direction. One particularly important parameter associated with any transistor is its transition frequency f_T, which applies when the device's current gain $G^{SC}{}_i$ is unity.

By definition:

$$G^{SC}{}_i = \left|\frac{i_C}{i_B}\right| = \left|\frac{g_m V_\pi}{i_B}\right| \tag{2.10}$$

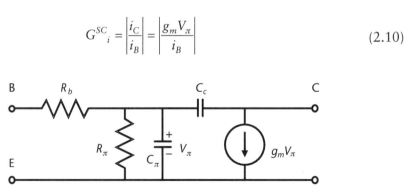

Figure 2.17 Small-signal hybrid-π equivalent circuit for a BJT.

Neglecting any effects due to the R_π shunt resistance, i_B/V_π is the susceptance of the capacitance $B_{gs} = 2\pi f C_{gs}$.

Substituting this into (2.10), setting $G^{SC}{}_i = 1$, and solving for frequency f_T yield:

$$f_T = \frac{g_m}{2\pi C_\pi} \tag{2.11}$$

For example, a BJT has transconductance $(g_m) = 1$ S and input capacitance $(C_\pi) = 8$ pF. What is the (upper limit) transition frequency for this device?

Substituting directly into (2.9), it is found that $f_T \approx 20$ GHz.

This is not particularly high when it is appreciated that many amplifiers implementing transistors will be required to operate at frequencies up and beyond 20 GHz. It is usually impractical to try increasing g_m and therefore attention must be paid to decreasing C_π. In practice, the equivalent circuit will include extra elements (resistances and inductors) at all three terminals to account for connection-point parasitics.

The microwave capabilities of silicon BJTs are seriously limited due to: bandgap narrowing, emitter current crowding, base access resistance, and base transit time.

As a result of these limitations gallium arsenide heterojunction bipolar transistors (GaAs HBTs) have been introduced. These types of RF transistors offer good high-frequency performance, particularly regarding low noise behavior. The cross-section of a GaAlAs/GaAs HBT is shown in Figure 1.17.

The notations n^+ and p^+ refer to relatively highly-doped regions which therefore exhibit relatively high conductivity. To indicate an NPN GaAlAs/GaAs HBT in a circuit the same symbol that is shown in Figure 1.3(b) can be used.

These types of transistors exhibit very good (wideband) input and output matching capabilities on account of the dominantly resistive natures of the device's input and output impedances.

When implemented within an MMIC design using NPN GaAlAs/GaAs HBTs, the overall result is a notably low noise figure and correspondingly low phase noise. These features are discussed much more fully in Chapter 8, considering a wide variety of options. HBTs can also be used in low-power RF amplifiers [4].

2.5 MMICs and RFICs

Fully integrated microwave (or RF) ICs are widely available and some basic concepts relating to their manufacture are presented in Section 2.2.6.

The following commonly encountered terms are essentially interchangeable: MMIC or RFIC. There are some arguments regarding the strict applicability of each terminology but the outcome is largely academic and will not be followed up here. The abbreviation MMIC will be the main one used in this book.

Unlike integrated circuits designed for digital subsystems (ICs or chips), those designed for RF or microwave applications cannot generally only comprise the transistors, interconnected by conductor arrays. Instead a combination of semiconductor devices (diodes and transistors) and passive circuit elements (see Chapters

3 and 4) is required. The final MMIC may embody a specific microwave circuit function such as an amplifier or a mixer or it may comprise a much more comprehensive subsystem.

In most instances (although not exclusively), the more comprehensive MMICs tend to be those realized using either RF CMOS or SiGe BiCMOS technologies. Major manufacturers of these more comprehensive chips offer fully integrated subsystems that include several ADCs, several DACs, and many further subcircuits all on the same chip.

Specific MMIC design approaches require the initial choice as to the transistor technology that will be implemented. This is referred to as the process. This means that a specific manufacturer (a foundry) will, for example, operate a publicized GaAs pHEMT process or a GaN HEMT process.

Providers of EDA software, such as Ansys, Keysight Technologies, or NI AWR, offer well-established and regularly updated design software that handles MMIC design as well as hybrid RF circuit design suites.

References

[1] Sze, S. M., and K. K. Ng, *Physics of Semiconductor Devices*, 3rd ed., New York: John Wiley & Sons, 2006.

[2] Shur, M., "Physics of Semiconductor Devices," Upper Saddle River, NJ: Prentice Hall Inc., 1990.

[3] Pozar, D. M., *Microwave and RF Design of Wireless Systems*, New York: John Wiley & Sons, 2001.

[4] https://www.microwaves101.com/encyclopedias/microwave- semiconductor- processing#typical.

[5] Kingsley, N., and J. R. Guerci, *Radar RF Circuit Design*, Norwood, MA: Artech House, 2016.

[6] https://patents.google.com/patent/US8076228B2/en.

[7] Ghavidel, A., et al., "GaN Widening Possibilities for PAs," *IEEE Microwave Magazine*, June 2017, pp. 46–55.

CHAPTER 3

Passive RF Components

3.1 Introduction

No RF or microwave circuit can be designed using entirely active devices. Instead, a carefully selected combination of transistors, diodes, and passive components must be used in carefully designed combinations, forming complete circuits. Therefore, passive components (capacitors, inductors, resistors, and transmission lines) are all vital for successful RF or microwave design.

The great majority of passive components are built using combinations of metals and dielectrics (insulators), unlike active components that utilize semiconductors. This chapter focuses on a wide range of passive components required for RF/microwave design.

Circulators and isolators are omitted here because these types of passive components are rarely used in communications systems, except sometimes in measurement setups. Much of the material presented here is required to support various types of circuits and module designs, and are dealt with elsewhere in this book.

3.2 Discrete Passive RF Components

3.2.1 Capacitors

Miniaturized, the most basic form of parallel-plate capacitor is frequently used in RF/microwave circuits, particularly hybrid designs. The basic arrangement is shown in Figure 3.1.

The free-space permittivity is ε_0 (= 8.854×10^{-12} F.m^{-1}) and the relative permittivity of the dielectric is ε_r then the classical result for the capacitance is:

$$C = \varepsilon_0 \varepsilon_r A / d \qquad (3.1)$$

with the area A and separation d both expressed in meters the capacitance computes in farads. Equation (3.1) neglects the additional aspect of side field fringing, which is usually very small for relevant structures.

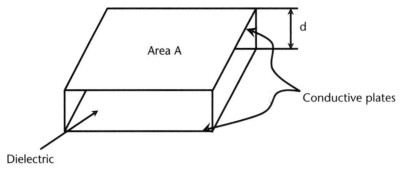

Figure 3.1 Basic concept of the parallel-plate capacitor.

Example 3.1

Calculate the area required to realize a capacitance of 10 pF, using a 0.2-mm-thick dielectric having relative permittivity 2.3. Rearrange (3.1) for area A and substitute the input data:

$A = 10 \times 10^{-12} \times 0.2 \times 10^{-3}/8.854 \times 10^{-12} \times 2.3$

Giving: $A = 0.098 \times 10^{-3}$ m² or $A = 98.2$ mm² (e.g., plate dimensions 9.91 mm × 9.91 mm or almost 1 square cm).

To reduce this area a higher permittivity dielectric filling could be used.

Example 3.2

Calculate the area required to realize a capacitance of 3 pF, using a 0.1-mm thickness of polyimide dielectric having relative permittivity 12.

Rearrange (3.1) for area A and substitute the input data:

$A = 3 \times 10^{-12} \times 0.1 \times 10^{-3}/8.854 \times 10^{-12} \times 12$

Giving $A = 0.00282 \times 10^{-3}$ m² or $A = 2.82$ mm² (e.g., plate dimensions 1.68 mm × 1.68 mm).

These data are just about compatible with realizing this capacitance using a polyimide dielectric in a MMIC environment.

All types of practical capacitor are imperfect. As well as the basic capacitance (C) there exists series parasitic inductance (L_s, due to lead connections), equivalent series resistance (R_s) representing losses in the dielectric and the metal electrodes and parallel parasitic capacitance (C_p) between lead connections. These effects lead to the equivalent circuit shown in Figure 3.2.

The quality factor (Q-factor) is an important characteristic for many types of passive components. For a capacitor, referring to values indicated in Figure 3.2, the Q-factor is given by the expression

$$Q = \frac{1}{2\pi f C R_s} \qquad (3.2)$$

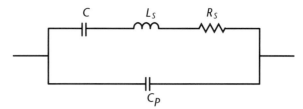

Figure 3.2 Equivalent circuit of a practical capacitor

Example 3.3

Calculate the Q-factor applying to a 3-pF capacitor having a series resistance of 0.5Ω, in a circuit operating at 20 GHz.

Using (3.2) and substituting, gives the answer as just 5.3, which is a very low Q-factor.

The only way to increase the Q-factor would be to select metal electrodes and/or metallic attachment technology that significantly reduces the series resistance R_s. For example, if R_s could be decreased to 0.2Ω, then the Q-factor would increase to 13.3, which is more than double the original value.

Regarding on-chip (MMIC-based) capacitors, there are three primary forms as follows:

- Metal-oxide-metal (MOM), realized with the interconnect metallization;
- Metal-oxide-semiconductor (MOS), comprising an MOS transistor;
- Semiconductor junction, realized either using the capacitance of a reverse biased PN junction (bipolar) or a Schottky barrier.

MOM capacitors typically enable values of some fF/μm^2 (femtofarads per square micron) to be realized. The Q-factors are reasonably high due to the low losses (metals: M) but the tolerance is poor because it is difficult to control the parameters associated with the oxide (O). MOS capacitors yield higher values, around 1 to 5 fF/μm^2 although losses are high (due to the use of high-resistivity semiconductor) leading to low-Q. Strong voltage and temperature dependencies also prevail. Semiconductor junction capacitors allow moderate values, but again strong voltage and temperature dependencies exist.

3.2.2 Inductors

Inductors represent significant passive components that are often required in RF/microwave circuits. These types of components are used in many circuit designs, usually in combination with other elements such as capacitors, or resistors, or sections of transmission lines. Because they effectively block RF while passing DC, inductors are also important in the realization of bias networks applying to active devices such as transistors.

Inductors having values up to about 10 nH can be fabricated on-chip. Where values above approximately 10 nH are required, these would occupy excessive die area and as a result all or most of such inductances must be realized off-chip (i.e., on the circuit board or card). Very small inductances (around the 0.5-nH to

1-nH range) may be realized using bond wires or tapes (including air bridges) on a MMIC or RFIC.

An example of a spiral inductor, implemented on-chip, is shown in Figure 3.3.

In practice, the air bridge could be supported mechanically by a thin layer of polyimide beneath it and above the spiral sections. Wheeler [1] provides the following formula for the inductance L of this structure:

$$L \approx \frac{9.4\mu_0 n^2 a^2}{11d - 7a} \quad (3.3)$$

Dimensions a and d are defined in Figure 3.3, n is the number of turns within the spiral, and μ_0 is the free-space permeability ($4\pi \times 10^{-7}$).

Although this formula (3.3) was derived for circular coils it has been verified as remaining accurate within 5% for square spirals (tested using the SONNET electromagnetic simulation package).

For example, if $n = 3$ (turns), $a = 200$, and $d = 400$ then substituting into (3.3) yields the inductance value of 1.42 nH.

It is typical for this type of inductor structure to provide inductance values of several nH, as required for many types of MMIC, RFIC, or hybrid microwave circuit.

Practical inductors inherently possess resistance due to the metal structure and parasitic capacitance across the terminals. These effects are represented as series resistance R_s and parallel capacitance C_p in the equivalent circuit shown in Figure 3.4.

The quality factor (Q-factor) is an important characteristic of an inductor, and referring to parameters shown in Figure 3.4, the Q-factor is given by

$$Q_L = \frac{2\pi f L}{R_s} \quad (3.4)$$

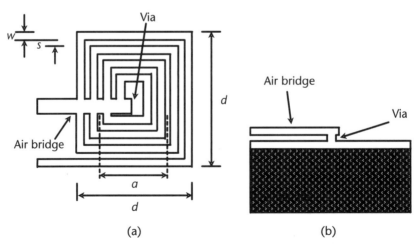

Figure 3.3 Example of a practical spiral inductor: (a) plan view and (b) side elevation.

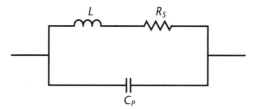

Figure 3.4 Equivalent circuit of a practical inductor.

where the frequency f is in gigahertz and the inductance L is in nanohenries (nH) the orders of magnitude cancel and the data can be substituted directly into (3.4) (i.e., the frequency has 10^9 associated with it, while the inductor has 10^{-9}). For example, if the 1.42-nH inductor cited above has a series resistance of 0.5Ω and operates at a frequency of 14 GHz, then the Q-factor calculates to 249.8. This represents a very good Q-factor, which would indicate a potentially effective inductor that could be used, for example, in a resonant circuit.

3.2.3 Resistors

These types of components are implemented as sparsely as possible in RF/microwave applications because they reduce signal levels and resistors increase the unwanted noise. However, in circuits such as Lange couplers and Wilkinson dividers (both considered in Chapter 4), nominally 50-Ω or 100-Ω resistors are required.

The practical realization of resistors depends critically on whether the choice is hybrid or monolithic (MMIC). In the case of a hybrid circuit (also some MMICs) it is best to select film-based resistors because this approach provides the following advantages:

- The manufacturing technology is highly compatible with the generation of the conductor circuit pattern in most instances.
- Resistor value control is obtained by laser trimming, which means that very fine tolerances are available.
- The material structure is particularly homogeneous, which leads to low flicker noise.

Evaporated nichrome or sichrome is frequently used as thin-film resistor materials while combinations of ceramic and metal (cermet) materials are widely implemented for thick-film resistor processes. Since a typical nichrome resistor film will have a resistivity of around 100Ω per square, then to realize a 50-Ω resistor demands effectively two squares in parallel (e.g., a structure 2 mm in length and 4 mm in width for a hybrid circuit). Alternatively, for a MMIC realization, the resistor could be 200 μm in length and 400 μm in width. It is also noted that the power-handling capability of the hybrid resistor (being dimensionally larger) would be superior to that of the MMIC version.

Resistors can be manufactured on-chip (MMIC) by implementing one of the following approaches:

- Thin metal conductor lines (interconnect);
- Polysilicon (in silicon RFICs);
- Specially doped semiconductor regions.

Some basic information is provided in Table 3.1 concerning various on-chip resistor options. These data were obtained from [2].

For all the options cited in Table 3.1, tolerances are poor and temperature coefficients are high.

The ideal design aims to, in all instances, include very good linearity, acceptable tolerance control, low parasitic capacitance, and a sufficiently small temperature coefficient of resistance. In practice, however, most of these aims cannot be met at a level compatible with circuit design requirements and as a result there is almost always a preference toward relying on resistor ratios rather than absolute values.

Practical resistors inherently possess parasitic inductance as well as parasitic capacitance across the terminals. These effects are represented as series inductance and parallel capacitance in the equivalent circuit shown in Figure 3.5.

This type of equivalent circuit must be introduced within any circuit design. As operating frequencies increase, so the effects of all parasitic elements are increased.

3.3 RF Transmission Lines

In contrast with lumped discrete passive RF components, transmission lines are distributive in nature (i.e., capacitive, inductive, and resistive effects are distributed in a continuous manner along any transmission line). These passive structures are important for interconnections between modules as well as forming significant types of circuit elements. A wide variety of forms of RF transmission lines exists and in the following sections some specific design details are given. Inherent losses are always present and these are very important to take into account. Losses comprise conductor losses in the metal fabric, dielectric attenuation, dielectric conduction (e.g., in high-resistivity silicon), and radiation.

Conductor losses are proportional to the square root of the frequency ($\sqrt{f}$) while losses in the dielectrics ($\tan \delta$) are directly proportional to the frequency.

All types of losses are covered in detail in [2]. Radiation losses are expanded upon in Chapter 7.

Table 3.1 Resistivities Associated with Some On-chip Resistors

Type	Resitivity (Ω per square)
Metal interconnect	0.01
Polysilicon (silicided)	5–10
Polysilicon (unsilicided)	50–100
Source-drain diffusion (FET)	25–200
MOS transistor	1,000–10,000

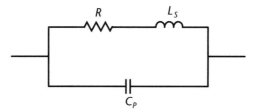

Figure 3.5 Equivalent circuit of a practical resistor.

3.3.1 Coaxial Lines

Invented in 1880 by English engineer and mathematician Oliver Heaviside, coaxial transmission lines predate almost any other form of RF transmission technology. About the only exception is conductor pairs, but these radiate energy to an unacceptable extent. The general (axial) cross section of a coaxial cable is shown in Figure 3.6.

This structure is termed transverse electromagnetic (TEM) because of the single-mode orthogonal nature of the electric and magnetic fields that comprise the electromagnetic wave progressing along this type of line.

From the electrical design viewpoint the critical parameters are:

- The inside diameter of the outer metal sheath (D);
- The diameter of the center conductor (d);
- The relative permittivity of the dielectric filling (ε_r).

The conductors are always nonferrous (i.e., usually either copper or aluminum) and the dielectric filling generally comprises a low-loss polymer. Using the dimensional parameters and the permittivity, the important design equations for the characteristic impedance (Z_0), propagation velocity (v_p), and the guided wavelength (λ_g) of the line are now presented. First, the characteristic impedance, assuming a loss-free line (most are very close to this situation):

$$Z_0 = \frac{60}{\sqrt{\varepsilon_r}} \ln \frac{D}{d} \; \Omega \tag{3.5}$$

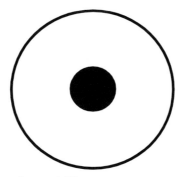

Figure 3.6 Cross sectional view of a coaxial line.

Note that this is a real number meaning this impedance is actually a resistance. It probably ought to have been termed characteristic resistance, but the term characteristic impedance has become the de facto standard.

Also, in the majority of instances, the characteristic impedance of microwave-grade coax is set at 50Ω. The arithmetic mean of 30Ω (for best power-handling capability) and 77Ω (for lowest loss) is 53.5Ω and the geometric mean is 48Ω. Thus, the somewhat rounded choice of 50Ω is a compromise between power handling capability and signal loss per unit length, for air dielectric. The 50-Ω standard is important throughout microwave design. Connecting a 50-Ω resistive load to the signal output end of a line is said to match the line, meaning there will be no reflections from that load; all the power will be absorbed.

In contrast, connecting a short circuit to the signal output end of a line will ensure that all the power will be reflected from that point back to the signal input. In practice, short circuits are far more reliable and precise than open circuits. This is because short circuits totally (electrically) blank off the end of the line, preventing any end fringing fields or radiation, whereas open circuits are subject to these undesirable effects.

The propagation velocity and guided wavelength are given by:

- Propagation velocity:

$$v_p = \frac{c}{\sqrt{\varepsilon_r}} \text{ m/s} \tag{3.6}$$

- Guided wavelength:

$$\lambda_g = \frac{c}{f\sqrt{\varepsilon_r}} \text{ m} \tag{3.7}$$

The expressions for characteristic impedance, propagation velocity, and guided wavelength are very important in general because similar design equations for other practical transmission lines are also presented in this chapter.

Some coaxial lines are commercially usable at frequencies through X-band (i.e., to 12 GHz) and beyond. Precision coaxial lines for measurement purposes can be used at frequencies above 60 GHz.

However, coax is not compatible with the demands of planar circuits, and as a result, other transmission structures have been developed. Planar circuit designs are fundamental to most RF/microwave technology.

3.3.2 Microstrip

One major disadvantage with pure TEM transmission lines (such as coaxial lines) is the fact that components as input parts cannot be attached without serious difficulties. Almost all MMICs and discrete semiconductors include terminals that require grounding and to only one ground plane. Furthermore, such components would have to be extremely thin; otherwise, the electrical interruptions within the sandwiched layers would be excessive and would lead to transmission and reliability issues.

3.3 RF Transmission Lines

During the 1960s, it was appreciated that simply by flattening out coax it is possible to conceive an open structure onto which external components can be readily attached. This structure was soon dubbed microstrip and this technology is now extensively used in most hybrid circuits (HMICs) and MMICs. The physical nature of microstrip is illustrated in Figure 3.7.

This type of structure can be realized on passive substrates or alternatively on the semiconductor surfaces associated with MMIC or RFIC technology. Like coax, the metal is usually copper (thinly gold plated) or aluminum, while the substrate is often either alumina or a polymer-based material where the design is to be using hybrid technology.

As well as the dimensions shown in Figure 3.7 (notably the width w), practical circuits also require a knowledge of microstrip physical lengths ℓ. Because microstrip behaves electromagnetically in a much more complex manner than coax, the design process is returned to later here.

As described above, coaxial lines support TEM electromagnetic fields. Coax is a totally enclosed structure with 100% homogenous electrical characteristics. Contrastingly, microstrip is certainly not totally enclosed because the fields are partially in the substrate and also partially in the air space above the strip (or trace). This leads to a relatively complex transverse and longitudinal field distribution and the term quasi-TEM is used to characterize microstrip transmission.

A parameter known as effective microstrip permittivity (ε_{eff}) is important in microstrip design. The introduction of ε_{eff} is means that regular formulas for characteristic impedance, propagation velocity and guided wavelength can be used in the design process. Instead of ε_r, which is used in the case of coax, the quantity ε_{eff} is employed. In particular, the equations given above for the propagation velocity and guided wavelength in coax ([3.6] and [3.7]) become, for microstrip:

- Propagation velocity (microstrip):

$$v_p = \frac{c}{\sqrt{\varepsilon_{eff}}} \text{ m/s} \qquad (3.8)$$

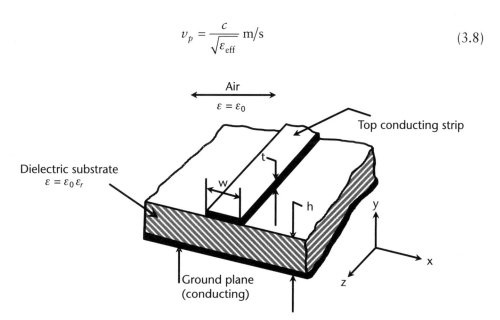

Figure 3.7 Three-dimensional (3-D) cross sectional view of microstrip.

- Guided wavelength (microstrip):

$$\lambda_g = \frac{c}{f\sqrt{\varepsilon_{eff}}} \text{ m} \tag{3.9}$$

Microstrip characteristic impedance Z_0 is fundamentally expressed as the root of the ratio L/C, exactly as for coax (see [3.5]), but the detailed expressions yielding L and C are relatively elaborate.

However, in terms of design procedures (termed synthesis here) for microstrip, the input parameters are characteristic impedance Z_0 (Ω) and choice of substrate (i.e., ε_r and h in mm or mils).

The required output data are the width w and physical length ℓ of the section of microstrip line.

Towards this aim, it is vital to evaluate the effective microstrip permittivity (ε_{eff}) because this quantity appears in every design formula.

3.3.2.1 Design at Relatively Low Frequencies

Over the years, the microstrip structure has been subjected to extensive DC-based analysis and synthesis. Although fundamentally imperfect, this static-TEM approach works well for fairly accurate design up to low to moderate microwave frequencies, depending on the detailed nature of the structure. In most instances, the static-TEM approach provides reasonably accurate output data for designs well into X-band.

First, an important comment regarding the limits of the effective microstrip permittivity ε_{eff}. The lowest relative permittivity in any relevant environment is 1 (the relative permittivity of air or space) and the highest value is , which is the relative permittivity of the substrate. Therefore, for any microstrip, the following inequality is true and important to remember: $1 \leq \varepsilon_{eff} \leq \varepsilon_r$.

For any microstrip therefore, the calculated (or measured) ε_{eff} must always be greater than or equal to 1 but less than the permittivity ε_r of the substrate.

For most practical purposes, the width w is normalized to the substrate thickness (height) h, that is, w/h. More complete details are available [2], but approximate design procedures are given here. It is necessary to bear in mind that relatively narrow strips are associated with relatively high characteristic impedances while the opposite applies to relatively wide strips.

First and foremost, it is essential to determine the w/h ratio and the approximate graphical technique originally due to Presser [3] is given here.

Presser's approximate design procedure begins with defining a quantity termed the microstrip filling factor q, which is a measure of the extent to which the substrate fills the entirety of the system, that is, q is determined by the permittivity of the substrate and the aspect ratio w/h. ε_{eff} and ε_r are interrelated by the following expression:

$$\varepsilon_{eff} = 1 + q(\varepsilon_r - 1) \tag{3.10}$$

Analytically, the details are elaborate but a consolidation of the main curves given as Figure 6.10 in [2] has been developed (Figure 3.8).

It is significant to observe that q becomes asymptotic to the value 0.75 as w/h trends towards infinity (in practice, when w/h exceeds approximately 4).

An approximate expression that fairly closely fits the data in Figure 3.8 is

$$q = 0.668(w/h)^{0.09} \tag{3.11}$$

This expression fits for q within a few percentage accuracy over the range: $0.58 \leq q \leq 0.75$.

Equation (3.11) is also restricted to the substrate permittivity range: $6 \leq \varepsilon_r \leq 40$.

For applications where the substrate permittivity is in the region of 2 to 3, around 3% should be added to the value of q obtained from Figure 3.8 or (3.11).

Presser's design procedure then is:

- Step 1: Make the initial (very rough) assumption $\varepsilon_{eff} \cong \varepsilon_r$ to obtain starting values.
- Step 2: Calculate the initial value of the air-spaced characteristic impedance, using:

$$Z_{01} \cong \sqrt{\varepsilon_r} Z_0 \tag{3.12}$$

- Step 3: Assume an initial approximate value for w/h (e.g., perhaps 1.0 for 50Ω, 1.5 for 25Ω, 0.5 for 75Ω, and so on) and apply this value to Figure 3.8 or substitute into (3.11) to determine the initial q.
- Step 4: Substitute this value of q into (3.10) to obtain the updated ε_{eff}.

Repeat steps 1 through 4, using updated values of ε_{eff} instead of ε_r in (3.12) until convergence is obtained i.e. the final values of ε_{eff}, w/h (hence, the width w because the substrate thickness h is known) and q are then all known. Finally, the guided wavelength λ_g is calculated using (3.9), directly in millimeters if the velocity of light c is written as 300 and the frequency is substituted in gigahertz.

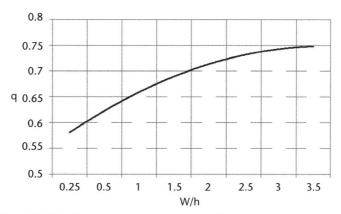

Figure 3.8 Microstrip filling factor q versus aspect ratio w/h.

As remarked above, this design procedure (or any other static-TEM design approach) works well for most microstrip transmission lines for frequencies up to and just into X-band (typically around 9 GHz).

3.3.2.2 Design for Higher-Frequency Operation

There is a fundamental electromagnetic field mechanism always operating in microstrip (also in other quasi-TEM transmission lines) known as dispersion and this significantly impacts design at frequencies lying in the higher X-band and beyond. Dispersion is caused by a mixing of the field modes, which is a phenomenon operating in all quasi-TEM transmission lines. This mixing intensifies as the operating frequency increases. An immediate effect of this dispersion is a distinct nonlinearity of the phase coefficient versus frequency relationship. For microstrip, this translates to ε_{eff} now actually itself being frequency-dependent so that this quantity must now be denoted $\varepsilon_{eff}(f)$.

Beginning in the late 1960s, many analyses of microstrip have been undertaken and published, including full-wave analyses accounting for dispersion. Also, various researchers have published formulas (or groups of formulas) providing closed-form expressions for $\varepsilon_{eff}(f)$.

Every formula relating to the design of microstrip at high-frequencies relies on already possessing the value of ε_{eff} (i.e., the low-frequency limit value). Also, some dispersion formulas require the characteristic impedance Z_0, which would be known at the outset from circuit design data. In general therefore the input data required are: ε_{eff}, ε_r, h, f (GHz) sometimes also w/h and Z_0.

It is important to recall the permittivity limits that exist in any microstrip transmission line structure, namely, $\varepsilon_{eff} \leq \varepsilon_{eff}(f) \leq \varepsilon_r$.

The lowest limit for $\varepsilon_{eff}(f)$ is the effective microstrip permittivity calculated under the static-TEM conditions, that is, ε_{eff}, described in Section 3.3.2.1. The highest conceivable limit for $\varepsilon_{eff}(f)$ is the substrate permittivity ε_r which would only apply if essentially all the space surrounding the strip were filled with substrate material. As the frequency is swept downwards and upwards, $\varepsilon_{eff}(f)$ approaches these limits asymptotically. Elsewhere, the function is mathematically well behaved, forming a smooth curve as shown in Figure 3.9.

General formulas should therefore guarantee these limits as well as the smooth trend with frequency. The pattern of formula that maintains these requirements is

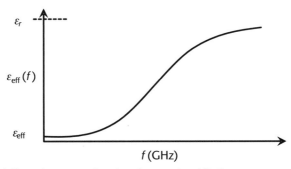

Figure 3.9 General dispersion curve showing the trend and limits.

3.3 RF Transmission Lines

$$\varepsilon_{\text{eff}}(f) = \varepsilon_r - \frac{\varepsilon_r - \varepsilon_{\text{eff}}}{1 + P(f)} \tag{3.13}$$

Most dispersion formulas focus on the polynomial function of frequency $P(f)$.

For practical design purposes, two formulas are recommended where the final choice depends on the specific microstrip specifications. The first expression was originally due to Edwards and Owens and is expanded upon in [2] being suitable for the design of microstrips using substrates having permittivities in the range

$$2 \le \varepsilon_r \le 10$$

thicknesses (h) in the range

$$0.4 \le h \le 0.8$$

and frequencies up to 20 GHz.

The formula is

$$\varepsilon_{\text{eff}}(f) = \varepsilon_r \frac{\varepsilon_r - \varepsilon_{\text{eff}}}{1 + \left(\frac{h}{Z_0}\right)^{1.33} (0.43 f^2 - 0.009 f^3)} \tag{3.14}$$

In which h is substituted directly in millimeters and f is substituted directly in gigahertz.

As an example, consider a 29-Ω microstrip manufactured on a 0.6-mm alumina substrate for which ε_{eff} calculates as 7.2. The relative permittivity ε_r of the alumina is 9.8. What is the guided wavelength at 12 GHz?

First, substitute the known parameters into (3.14),

$$\varepsilon_{\text{eff}}(12) = 9.8 - \frac{9.8 - 7.2}{1 + \left(\frac{0.6}{29}\right)^{1.33} (0.43 \times 12^2 - 0.009 \times 12^3)}$$

that is,

$$\varepsilon_{\text{eff}}(12) = 9.8 - \frac{2.6}{1 + 0.00575 \times (61.92 - 15.55)}$$

or

$$\varepsilon_{\text{eff}}(12) = 7.75$$

which is 7.6% higher than the low frequency value of 7.2.
Using (3.9), the guided wavelength is calculated

$$\lambda_g = \frac{c}{f\sqrt{\varepsilon_{\text{eff}}}}$$

Substituting $c = 300$ and the 12-GHz frequency directly in gigahertz:

$$\lambda_g = \frac{300}{12\sqrt{7.74}}$$

that is,

$$\lambda_g = 8.980 \text{ mm}$$

which is 3.4% below the value of 9.32 mm using just ε_{eff}, which would lead to a significant error in circuit design.

Clearly, at higher frequencies, the error increases significantly.

However, Edwards and Owens' formula will not work as frequencies exceed 20 GHz and/or where substrates are thinner (less than about 0.4 mm). Therefore, for these conditions, a new expression is required and several are available in the literature. All published expressions are relatively complex and cumbersome to apply and as a result a new formula (of limited applicability) has been sought. Reference is made to Edwards and Steer's book [2], specifically Figure 7.19 on page 181, which provides a family of dispersion curves relating to microstrip on a semi-insulating GaAs substrate (MMIC) having the following characteristics:

$$\varepsilon_r = 13, \ h = 0.127 \text{ mm}, \ w = 0.254 \text{ mm}$$

The following observations apply to this family of curves:

1. Those curves relating to the simulation and also to Kirschning and Jansen match very closely.
2. Over the frequency range of 0 to 40 GHz, all the results exhibit an approximately square law pattern.

Therefore a straightforward quadratic equation was developed by curve-fitting to the mid-points between the Kirschning and Jansen and simulation data. The general result is

$$\varepsilon_{\text{eff}}(f) = \varepsilon_{\text{eff}} + Mf^2 - Nf \tag{3.15}$$

where Mf^2 is the main frequency-dependent term and Nf accounts (to a limited extent) for the initial roll-off at very high frequencies. The frequency f is directly substituted in gigahertz.

Setting $M = 0.000374$ and $N = 0.01M$ yields a good fit across the range of frequencies up to at least around 80 GHz and this expression works well for substrates having permittivities in the range

$$10.1 \leq \varepsilon_r \leq 15$$

and thicknesses (h) in the range

$$0.1 \leq h \leq 0.39$$

A word of warning is however in order here. Equation (3.15) becomes inapplicable at extremely high frequencies because its use could easily lead to $\varepsilon_{eff}(f)$ becoming $\geq \varepsilon_r$, which is unacceptable. Over the ranges specified above this equation works well and reliably.

In overall guidance, if you are designing a hybrid RF circuit operating at a top frequency below about 20 GHz, then use (3.14). If you are designing a MMIC or an RFIC operating at a top frequency anywhere between 20 GHz and about 80 GHz, then use (3.15).

3.3.2.3 Some Important Discontinuities

Practical circuits do not just comprise straight sections of regular microstrip lines. Instead, necessarily discontinuities arise and many of these can be dealt with by compensation techniques. This section deals with a few of the most commonly encountered microstrip discontinuities.

First, for the microstrip open end, superficially any microstrip that abruptly stops on the surface of its substrate might be thought of as just that: an immediate stop with a clean open circuit. However, this is never the case because fringing fields continue to exist past the physically open circuit end and this is called the open end-effect. It is relatively easy to allow for this by equivalently adding a very short section of line onto the end of the actual microstrip.

Second, for right-angled bends in microstrip, Figure 3.10 depicts a regular right-angled bend and a miter-compensated right-angled bend.

Several recommendations and expressions claiming optimum miter designs have been reported in the literature, but proportionately the outline shown in Figure 3.10(b) works well in practice.

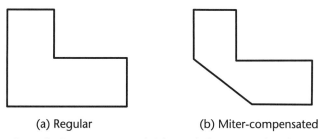

(a) Regular   (b) Miter-compensated

Figure 3.10 Regular and miter-compensated right-angled microstrip bends.

3.3.2.4 The Radial Stub

Instead of a conventional microstrip line, used as a stub for matching purposes, a radial stub is often implemented. The basic layout of this type of microstrip structure is shown in Figure 3.11.

Atwater [4] performed a computer-based analysis on this type of structure, and based on the results, he developed the following expression for the design radius r_2.

$$\log r_2 = A \log\left(\sqrt{\varepsilon_r} f_{so}\right) + B \log h + C \log r_1 + D \qquad (3.16)$$

In which the radii r_1 and r_2 and the substrate thickness h must all be input in meters, ε_r is the permittivity of the substrate material and f_{so} is the operating (center) frequency that must be directly input in gigahertz. In Atwater's system, only two angles (α) are allowed: 60° and 90°. Table 3.2 (directly from Atwater's paper) provides the parameters A, B, C, and D for use with (3.16).

In a numerical example, consider a 60° radial stub for which the following (input) parameters apply: operating (center) frequency f_{so} = 20 GHz, $r_1 = h = 0.5$ mm, and the substrate is alumina with permittivity 9. The radius r_2 must be determined.

For the solution, use the equation in conjunction with the appropriate parameters given in Table 3.2 (with $\alpha = 60°$). The numerical results for each term in the equation are:

$$\log r_2 = -0.8232 \log\left(\sqrt{9} \times 20\right) + 0.0572 \log\left(5 \times 10^{-4}\right)$$
$$+ 0.1169 \log\left(5 \times 10^{-4}\right) - 0.8082$$

or

$$\log r_2 = -1.4638 + 0.04 - 02288 + 0.0817$$
$$- 0.4676 - 0.0802$$

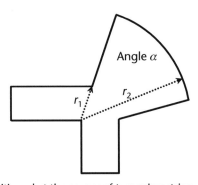

Figure 3.11 Radial stub positioned at the corner of two microstrips.

Table 3.2 Radial Stub Design Parameters

α	A	B	C	D
60°	−0.8232	0.0572	0.1169	−0.8082
90°	−0.8510	0.0614	0.0877	−0.8595

Source: [4].

which leads to: $\log r_2 = -2.8467$, that is, $r_2 = 0.001423$ meters or $r_2 = 1.423$ mm, which will almost certainly be truncated to three significant figures (i.e., $r_2 = 1.42$ mm).

Alternatively, a freely available computer program can be used for this determination [5]. Care should be taken with this program because the input dimensions can be selected either in millimeters or inches. This procedure works well for substrates having permittivities in the range $2 \leq \varepsilon_r \leq 15$ and for frequencies (f) in the range $0.3 \leq f_{so} \leq 30$ GHz.

Operation at higher frequencies and with different substrates should be carefully examined before going ahead with any new design.

Filter circuits often comprise radial stubs arranged in pairs, one connected to each side of the main microstrip line. A pair of radial stubs connected in this fashion is termed a butterfly stub or a bowtie stub. A typical bandpass filter designed using this approach will have five stages.

All the above technologies involving radial and butterfly stubs exhibit significantly wider bandwidths than conventional (linear) microstrips.

3.3.2.5 Limitations of Microstrip

Microstrip is limited in terms of maximum operating frequency and also its inherent Q-factor [2]. Operating frequency limits are set by:

- The lowest-order transverse microstrip resonance;
- Onset of the lowest-order transverse magnetic mode (i.e., TM mode).

Refer back to Figure 3.7 and consider the effect of the width (w). Fringing fields extend each side of the strip, approximately to a distance $d = 0.2h$ in each case. Therefore, at a sufficiently high frequency, a half-wavelength can be supported transversely across the distance $w + 4h$. The frequency associated with this resonance is the first frequency limitation.

The lowest-order TM mode is more complex to analyze and eigenvalues are required [2]. The result for the frequency of the onset of the lowest-order TM mode is given by

$$f_{TEM,1} = \frac{c}{2\sqrt{2}h\sqrt{\varepsilon_r - 1}} \quad (3.17)$$

Q-factor is degraded by the losses in the line, referred to early in Section 3.3. Microstrip Q-factors are relatively poor, usually amounting to a few hundred at best. This feature severely restricts the use of this transmission medium in oscillators (Chapter 12).

3.4 Coplanar Waveguide

Invented by a Japanese engineer named C. P. Wen, coplanar waveguide (CPW) represents a further technology choice for RF/microwave transmission. The most basic version of CPW simply comprises a group of three metal structures deposited on one side of a substrate: the main stripline as well as two side-coupled and grounded metallic regions. However, in most practical instances, a further ground plane is included on the reverse side of the substrate and this topology is termed grounded CPW (GCPW).

The basic structure of GCPW is illustrated in Figure 3.12 where the structure is assumed to be symmetrical with strip width w and equal longitudinal gaps having dimension s. The two (left and right) top-side conductors are ultimately grounded, theoretically at infinity. In this cross sectional diagram, all the black-filled regions are metal.

Notice how if the two top-side grounded regions are removed, this structure reverts to standard microstrip.

This structure has the following advantages over microstrip:

- Easier grounding of surface-mounted components;
- Lower fabrication costs;
- Greatly reduced dispersion (for small geometries such as MMICs);
- Decreased radiation losses;
- Couplers can be realized having higher directivities;
- These photolithographically defined structures have a relatively low dependence on substrate thickness.

The design of GCPW begins, like microstrip, with a knowledge of the substrate parameters (ε_r and h) as well as the characteristic impedance Z_0. Again as with microstrip an effective coplanar waveguide permittivity ε_{eff} is defined, which is used to determine the guided wavelength λ_g. Equation (3.9) applies, namely:

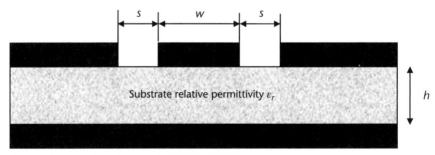

Figure 3.12 Cross section of grounded CPW (GCPW).

$$\lambda_g = \frac{c}{f\sqrt{\varepsilon_{\text{eff}}}} \text{ m}$$

where ε_{eff} is now the GCPW value. This value will always be considerably lower than for microstrip because a greater proportion of the fields are distributed in the air spaces (*s*) between each side of the strip and the top ground planes.

However, the fact that microstrip inherently binds the electromagnetic fields relatively tightly within the substrate plus the fact that a wealth of design experience exists means that microstrip technology dominates.

Also, microstrip is extensively used for designing the great majority of the circuits described in this book. All of this means that CPW (and GCPW) is selected in far fewer circuit designs than microstrip.

3.5 Substrate Integrated Waveguide

First introduced by Wu, Deslandes, and Cassivi in 2003, substrate integrated waveguide (SIW) is a relatively new form of transmission line that is gaining prominence for some RF and microwave applications [6]. The basic concept of SIW is illustrated in plan (top) view in Figure 3.13.

In Figure 3.13 the elements denoted • represent sets of plated-through vias forming two sides of a dielectric-filled rectangular waveguide. The structure is completed by the underlying ground plane and a top conducting plane, which covers the portion of the circuit existing between the two rows of vias. In this way, a localized dielectric-filled rectangular waveguide is formed.

A major advantage is the fact that this waveguide largely comprises metal elements, which leads to substantially decreased losses compared with microstrip or coplanar waveguide and hence higher Q-factors.

However the accurate and repeatable manufacture of SIW represents a major challenge. The vias must:

1. Be set close enough together to avoid energy leakage (i.e., the separation must be less that about $0.2\lambda_g$ at the highest frequency of operation).
2. Be accurately and automatically positioned within a fraction of a millimeter (for hybrid circuits) or within a few micrometers for MMICs or RFICs.

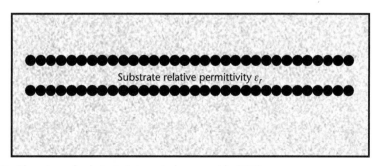

Figure 3.13 Plan view of substrate integrated waveguide (SIW).

References

[1] Wheeler, H., "Simple Inductance Formula for Radio Coils," *Proc. of the Institute of Radio Engineers*, Vol. 16, No. 10, October 1928, pp. 1398–1400.

[2] Edwards, T., and M. Steer, *Foundations for Microstrip Circuit Design*, 4th ed., New York: John Wiley & Sons, 2016.

[3] Presser, A., "RF Properties of Microstrip Lines," *MicroWaves*, Vol. 7, March 1968, pp. 53–55.

[4] Atwater, H., "The Design of the Radial Line Stub: A Useful Microstrip Circuit Element," *Microwave Journal*, November 1985, pp. 149–156.

[5] http://flambda.com/stub/stub.php.

[6] Rayas-Sanchez, J. E., and V. Gutierrez-Ayala, "A General EM-Based Design Procedure for Single-Layer Substrate Integrated Waveguide Interconnects with Microstrip Transitions," *IEEE MTT-S Int. Microwave Symp. Dig.*, Atlanta, GA, June 2008, pp. 983–986.

CHAPTER 4

Passive RF Circuit Elements

4.1 Introduction

This chapter focuses on couplers, power splitters, power combiners, and baluns. The material presented here in many respects builds on aspects of Chapter 3. In general, these kinds of circuit elements are frequently chosen for implementation in various types of RF and microwave circuits. In this chapter, the fundamentals of couplers are considered in detail. This is followed by detailed considerations of the Lange coupler, Wilkinson power dividers, and finally baluns (several details are provided in [1]. There are many other types of couplers and dividers but major examples are studied in detail here.

4.2 Fundamentals of Directional Couplers

In RF and microwave engineering, it is often necessary to separate and direct signal energy into two or more ports. For this purpose, some form of coupler is required and such a device represents a basic passive RF circuit element.

Basic concepts (and symbols) relating to directional couplers are shown in Figure 4.1.

The two output ports comprise P_2 and P_3. Assuming a 3-dB coupler, then it follows that $|S_{21}| = |S_{31}| = -3$ dB nominally, losses to be included as well because of attenuation within the coupler structure itself.

Where the output levels are unequal, the coupling factor C must be found using (4.1):

$$C = 10\log\frac{P_1}{P_3} = 20\log\frac{1}{|S_{31}|} \text{ dB} \quad (4.1)$$

in which P_1 is the signal power input at port 1 (in watts) and P_3 is the signal power output at port 3 (also in watts).

Ideally (i.e., for a perfect coupler), there would be zero output power from the (Isolated) port 4. A power detector connected to port 4 would read zero at

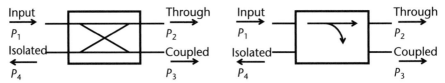

Figure 4.1 Equivalent symbols for directional couplers, showing the signal ports [1]. (© Artech House, 2016.)

maximum sensitivity, but in practice there would always be some power level appearing at the isolated port 4 and this fact leads to two parameters, directivity (*D*) and isolation (*I*), determining the electrical quality of any coupler. *D* relates the output power levels associated with the coupled port (3) and the isolated port (4), expressed as:

$$C = 10\log\frac{P_3}{P_4} = 20\log\left|\frac{S_{31}}{S_{41}}\right| \text{ dB} \qquad (4.2)$$

The isolation (*I*) is a measure of the signal power leakage from the input port (1) into the isolated port (4) and is expressed as:

$$C = 10\log\frac{P_1}{P_4} = 20\log\frac{1}{|S_{41}|} \text{ dB} \qquad (4.3)$$

In practice any coupler design should have both *D* and *I* as large as possible, certainly exceeding 10 dB.

The simplest physical realization of a coupler comprises two microstrip lines (Chapter 3) running parallel to each other on the surface of a substrate with a small coupling gap between these traces. However, this approach results in narrowband and highly nonlinear coupling and more sophisticated structures are therefore demanded.

4.3 The Lange Coupler

Invented in 1969 by Julius Lange, this coupler is an ingenious and remarkable device that is capable of highly broadband operation and can be manufactured in either hybrid circuit or MMIC format.

Basically, the Lange coupler comprises a four-port circuit having a central arrangement of tightly coupled, interleaved fingers. The overall length of this interdigital coupling section is nominally $\lambda_g/4$ (although the choice of λ_g is critical for broadband operation). This type of coupler is often referred to as a quadrature coupler, on account of the nominally 90° phase difference between its two outputs.

Lange's aim was to produce an almost octave-bandwidth coupler having a coupling factor of around −3 dB. In Lange's empirical design, true quadrature coupling over an octave is realized as a consequence of the interdigital coupling section which compensates for even-mode and odd-mode phase velocity dispersion over a wide range of frequencies. It is necessary to bond directly, via electrically short conductors, between transversely interposed fingers of the coupler. With this circuit structure, RF power is coupled optimally to the desired port, in addition to the remaining power still being fed to a direct port. A plan view of a four-finger Lange coupler is shown in Figure 4.2.

The $\lambda/4$ length is nominal in Figure 4.2. Observe that the input-to-direct output link (a DC connection) meanders through the center of the structure.

A suitable circuit symbol is indicated in Figure 4.3.

From the electrical viewpoint, the bonding wires must look as close as possible like short circuits or at least very small lumped discrete inductances. This means that their physical lengths ℓ_s must be kept as short as possible in accordance with the condition:

$$\ell_s < 0.2\lambda_{gh} \tag{4.4}$$

where λ_{gh} is the wavelength associated with the highest frequency of operation.

The relatively short finger elements are approximately $\lambda_{gh}/4$ at the highest frequency in the band, whereas the entire length of the through section of line within the central region of the coupler is approximately $\lambda_{gh}/4$ at the lowest frequency of the band. The bonding conductors ensure properly phased traveling-wave reinforcement for the coupler. These three facets lead to the excellent performance almost always obtained with Lange couplers. There are two further advantageous features:

- Phase shift usually better than ±2° on the 90° nominal value across the operating band;

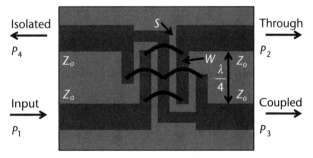

Figure 4.2 Plan view of a four-finger Lange coupler [1]. (© Artech House, 2016.)

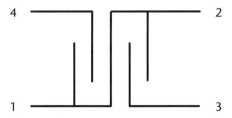

Figure 4.3 Lange coupler circuit symbol.

- Voltage standing-wave ratio (VSWR) usually less than 1.15 over an octave operating band.

Lange couplers designed for operation at frequencies above X-band require several bonding conductors placed close together in parallel to reduce the inductance. This results in an extra manufacturing expense (notably for hybrid circuit realization), since the wires must be delicately bonded in by hand or by precision computer-controlled machines.

Design starts with the following input data:

- The coupling factor C (as a linear factor, converted from decibels);
- The number of lines within the coupler;
- The characteristic impedance (Z_0) of the input and output microstrips.

It then proceeds as follows:

- *Step 1:* Determine the special coupled characteristic impedances. These are known as even-mode and odd-mode, resulting from the two extremes of field distributions between the strips.
- *Step 2:* From these characteristic impedances, calculate the widths (w) and spacings (s) between the coupled lines.

The strategy for calculating the widths is broadly similar to that outlined for single microstrip lines (Chapter 3).

Computing the physical lengths of the strips is relatively straightforward because the even and odd-mode propagation velocities are almost equal (ideally perfectly so), which means that the guided wavelengths λ_{gh} and λ_{gl} can both be calculated on the basis of one value of effective microstrip permittivity (ε_{eff}) using the basic single microstrip expression given as (3.9) of Chapter 3, that is:

$$\lambda_g = \frac{c}{f\sqrt{\varepsilon_{\text{eff}}}}$$

In practice, due to the fringing from the open ends, the physical lengths of each interior open-ended coupled section will be slightly shorter than calculated using the quarter-wavelength formulas.

Most EDA design environments include a routine for the design of a Lange coupler.

4.3.1 EM Structure

The microstrip Lange coupler model (MLANGE2) is shown in Figure 4.4. Equations are used to set variables for dimensions, so that the parameters of the coupler and the manifolds can be varied together.

The actual diagram presented in the NI AWR application includes extensive data relating to every segment of the layout shown in Figure 4.4.

4.4 Wilkinson Power Dividers

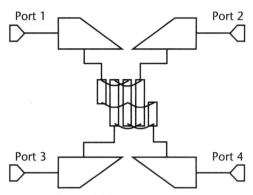

Figure 4.4 Microstrip Lange coupler model (NI AWR MLANGE2).

The final physical layout comprises relatively thin and proportionately long lines. It was too awkward a structural diagram to include in this book. The lines are 200 mil (5.08 mm) long and 5 mil (0.127 mm) wide.

Obviously, the physical dimensions of the fingers in Lange couplers decrease as the frequency increases and become extremely small at millimeter-wave frequencies, even on low permittivity substrates. In these instances, MMIC realization is important and the bond conductors are realized as conductive air bridges. Thin-film technology is regularly used with MMICs and either microstrip or coplanar waveguide (CPW) can be implemented. Figures 1.19 and 4.5 show E-band MMICs that include Lange couplers (Plextek RFI, private communication).

CPW pad connections are made available at both the inputs and the outputs of each chip shown in Figures 1.19 and 4.5. CPW is described in Chapter 3.

CPW has also been used as a basis for complete circuit designs implementing quadrature couplers [2].

4.4 Wilkinson Power Dividers

4.4.1 Introduction to Wilkinson Dividers

It is often necessary for signals to be separated into several paths or routes and power combiners or dividers provide this facility. Examples include parallel-combined power amplifiers or antenna arrays.

4.4.2 Equal-Split Wilkinson Dividers

Although the Wilkinson combiner may be used either strictly as a power combiner or a divider, it is most frequently used in divider mode. Two examples of equal-split Wilkinson dividers are shown in Figure 4.6.

The $\lambda/4$ lengths are nominal and are calculated at the center frequency for the system.

In each instance, the system impedance is 50Ω and the transmission lines (usually microstrip lines) perform the required impedance transformations.

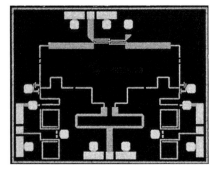

Figure 4.5 E-band MMIC implementing a Lange coupler at the output (top). (Courtesy of Plextek RFI.)

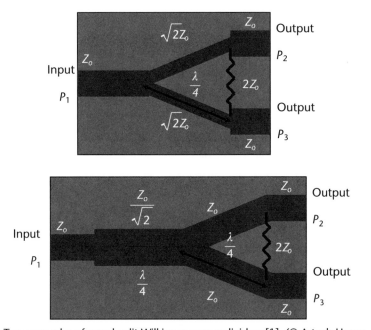

Figure 4.6 Two examples of equal-split Wilkinson power dividers [1]. (© Artech House, 2016).

A resistor ($2Z_0$) connected between the output ports ensures good isolation between them. In the case of the three-port, equal-split design all the ports are automatically matched to Z_0. This represents an important circuit advantage.

No power is absorbed by the isolation ($2Z_0$) resistor for the symmetrically split component of the output signal. When used as a combiner, the two signals presented at the junction must have exactly the same amplitude and phase to be combined. In practice (designing for a small error) this is difficult to achieve.

Where the system impedance is 50Ω, the value of the isolation resistor becomes 100Ω, which is readily achieved using the technologies described in Chapter 3.

4.4.3 Unequal-Split Wilkinson Dividers

Wilkinson power dividers having unequal output splits are also important and an example is shown in Figure 4.7.

4.4 Wilkinson Power Dividers

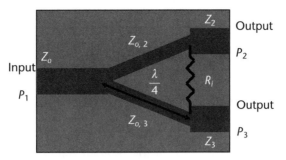

Figure 4.7 An unequal-split Wilkinson power divider [1]. (© Artech House, 2016.)

The ratio of the power levels at both output ports can be expressed as

$$r = \sqrt{\frac{P_3}{P_2}} \quad (4.5)$$

where all power levels (P_2 and P_3 here) are substituted in watts (W), not dBm.

The characteristic impedances of the two separated signal paths are given by:

$$Z_{0,2} = Z_0 \sqrt{r/(1+r^2)} \quad (4.6)$$

Most references repeat an error in the equation for $Z_{0,2}$ in that they all indicate a multiplication under the root sign. This should be a division and the correction is shown here.

$$Z_{0,3} = Z_0 \sqrt{\frac{1+r^2}{r^3}} \quad (4.7)$$

and the value R_i of the isolation resistor can be calculated from

$$R_i = Z_0 (r + 1/r) \quad (4.8)$$

The output port microstrip line characteristic impedances are determined using

$$Z_2 = Z_0 r \quad (4.9)$$

$$Z_3 = Z_0 / r \quad (4.10)$$

The impedances of the inner lines $Z_{0,2}$ and $Z_{0,3}$ are highly critical for the correct design of unequal-split Wilkinson power dividers. For example, if the system characteristic impedance is 50Ω and the power ratio r is, say, 8, then calculating using (4.6) and (4.7), one line must have a characteristic impedance $Z_{0,2}$ of 140.3Ω while the other line must have a characteristic impedance $Z_{0,3}$ of 17.5Ω. Obviously, in a 50-Ω system these impedances must be matched at their output ports.

The practical consequences of demanding microstrip lines with these characteristic impedances must be examined (Chapter 3).

The 0,3 line is manufacturable (albeit yielding a somewhat wide line), but the 0,2 line is essentially nonmanufacturable on any substrate because it would result in an exceptionally thin line, which would exhibit a highly uneven and unreliable impedance in practice and be subject to open-circuit breaks in the line.

A power ratio (r) of 4 leads to an 0,2 line with a characteristic impedance of 104Ω, which is just about manufacturable on most types of substrate. Therefore, in practice, it is best to avoid ratios exceeding about 3.

Figure 4.8 is a circuit diagram showing the unequal-split Wilkinson power divider with impedance matching on all three ports.

For this circuit the design equations are:

$$Z_1 = Z_0 \left(r / (1+r^2) \right)^{0.25} \tag{4.11}$$

$$Z_2 = Z_0 \cdot r^{0.75} \left(1+r^2 \right)^{0.25} \tag{4.12}$$

$$Z_3 = Z_0 \frac{\left(1+r^2\right)^{0.25}}{r^{1.25}} \tag{4.13}$$

$$R_i = Z_0 \frac{\left(1+r^2\right)}{r} \tag{4.14}$$

4.4.4 Multiport Equal-Split Wilkinson Dividers

It is also possible to extend the Wilkinson divider approach to multiple-output designs ($N \geq 4$). The layout of a four-way, equal split circuit is shown in Figure 4.9.

As a logical extension to the two-way dividers described above, these four-way circuits have their individual internal line impedances set to the square root of the

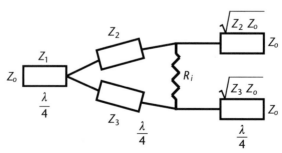

Figure 4.8 An unequal-split Wilkinson power divider with impedance matching on all three ports [1]. (© Artech House, 2016.)

4.4 Wilkinson Power Dividers

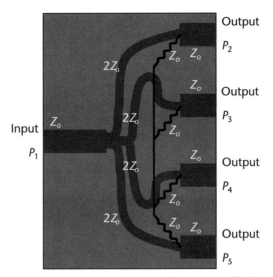

Figure 4.9 An equal-split, four-way Wilkinson power divider layout [1]. (©Artech House, 2016.)

number of outputs N (i.e., $\sqrt{N}$) multiplied by the system characteristic impedance. For the four-way divider shown in Figure 4.9 this impedance therefore equals $2\,Z_0$, which amounts to 100Ω where the system characteristic impedance is 50Ω. Once again, realizing a 100-Ω microstrip line is close to the limits in practice.

All internal line lengths are quarter-wavelength, all isolation resistors have the same value as the system impedance, and they each share a common terminal point. This configuration poses a substantial practical challenge because:

- The isolation resistors' common terminal must itself be isolated from the remainder of the circuit layout.
- It is important to achieve a high degree of overall isolation between the various elements of the circuitry.

Provided that the overall structure is physically small (occupying only some mm²), air bridges could be introduced to connect the ends of the resistors. This approach would be similar to that described for Lange couplers (Section 4.3). The issue of internal isolation remains, however. To some extent, the isolation effect is lessoned by manufacturing with as much symmetry as possible. Monolithically integrated dividers will best satisfy this requirement (MMICs).

EDA tools are widely available for the design of Wilkinson power dividers. For example, NI AWR provide the following (linked) two examples.

For the 4_Way_Power_Splitter design process, no input data are supplied, but (as for many designs) it is assumed that the substrate is 0.6 mm thick with a permittivity of 9.8.

The operating frequency band is from 10 to 18 GHz. Under EM Extraction – Core Schematic, the layout shown in Figure 4.10 is obtained for the basic two-way splitter.

The output response indicates approximately – 6 dB of insertion loss was obtained across the 10-GHz to 18-GHz band.

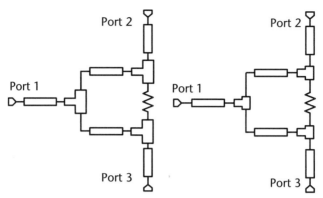

Figure 4.10 Layout of the basic two-way power splitter. (Based on figure in http://kb.aur.com/display/examples/4_Way_Power_splittler.)

In the same example, this two-way power splitter is then treated as one of the subelements forming the final four-way circuit as a combination of the basic two-way circuits with the result shown in Figure 4.11.

Output insertion losses are all cumulative on those cited above for the two-way splitters.

In Figure 4.11, each of the three two-way splitters is shown as a block with the single input to the LHS and the two outputs to the RHS. Also, all five ports are matched to 50Ω.

4.5 Baluns

On some occasions, it is necessary to convert between two distinctly differing circuit configurations: balanced transmission and unbalanced transmission. A pair of live conductors (neither being grounded) represents a balanced transmission structure while contrastingly a single live conductor above a ground represents an unbalanced transmission structure. Coax and microstrip are both good examples of unbalanced transmission structures.

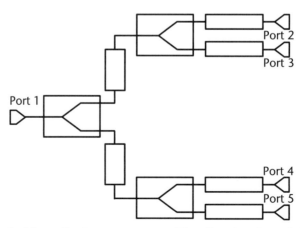

Figure 4.11 Layout of the entire four-way power splitter. (Based on figure in http://kb.aur.com/display/examples/4_Way_Power_splittler.)

4.5 Baluns

A type of passive circuit element termed a balun, invented by Marchand in 1944, provides the balanced-to-unbalanced conversion function. The basic principle of a balun is indicated in Figure 4.12.

This is a reciprocal circuit in that it can be used either way. Sometimes, when transforming from unbalanced to balanced, the circuit is termed an unbal and this terminology is occasionally used. At microwave frequencies baluns can be realized using 180° hybrid junctions.

Microstrip realizations of baluns are used in antennas, Class B push-pull amplifiers, differential RF paths on RFICs, mixers, and multipliers where balanced transmission line structures are needed to feed certain sections of the circuit, whereas unbalanced lines are then required to form the remainder of the circuit.

Baluns have been designed for both monolithic and hybrid integrated implementation.

Two practical balun layouts are shown in Figure 4.13. The top layout comprises a quarter-wave edge-coupled design, while the lower diagram is the configuration originally conceived by Marchand.

Although the edge-coupled line version is a relatively short structure ($\lambda/4$), it requires a series of bond wires (or air bridges). The Marchand balun is twice the length of the edge-coupled line version but avoids any requirement for bond wires or air bridges.

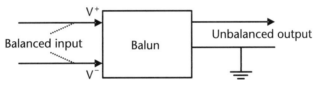

Figure 4.12 Basic principle of a balun.

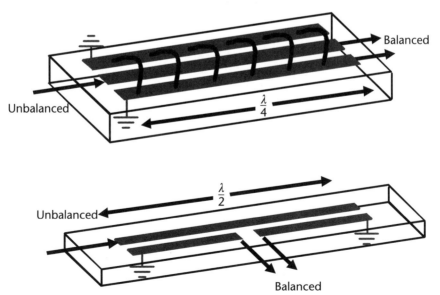

Figure 4.13 Two balun layouts (not to scale): edge-coupled line version (top), and Marchand realization (bottom) [1]. (© Artech House, 2016.).

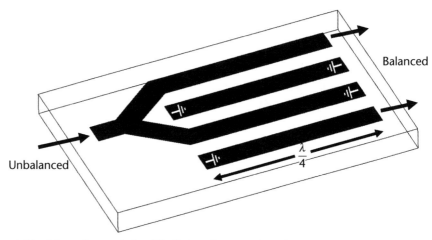

Figure 4.14 A four-element balun [1]. (© Artech House, 2016.)

The layout of a four-element balun is shown in Figure 4.14.

This balun (Figure 4.14) requires three short circuits to be realized in the structure. The accurate positioning of these is a significant challenge whether the circuit is implemented in hybrid or MMIC technology.

Baluns can be designed to enable a flat amplitude (within 1 dB) over a bandwidth of 6 to 18 GHz, with 180° (±2°) phase matching maintained between the output lines.

In some designs tight coupling is achievable by means of introducing a relatively thin dielectric layer on the surface of the circuit. In a MMIC implementation, this can be achieved with a 2-μm-thin layer of silicon nitride on the surface of the chip.

References

[1] Kingsley, N., and J. R. Guerci, *Radar RF Circuit Design*, Norwood, MA: Artech House, 2016.

[2] Kuo, C. -Y., et al., "A Compact 20-35 GHz Quadrature Coupler Using 0.15 μm pHEMT Coplanar Waveguide Technology," *2008 China-Japan Joint Microwave Conference*, September 10–12, 2008, pp. 636–639.

CHAPTER 5

Switches, Attenuators, and Digital Circuits

5.1 Introduction

High-speed switches, attenuators, and digital circuits are all increasingly important in many RF and microwave systems.

All of these types of circuits are considered in this chapter. However, while switches and attenuators are dealt with in considerable detail, no design details are provided for digital circuits because this would be beyond the scope of this text. Reference [1] is recommended for further details on digital circuits.

5.2 Solid State RF Switches

5.2.1 Some Overall Aspects

In this section, the fundamentals, operation, and technologies associated with solid-state RF switches are outlined. Basically, switches are needed in almost all electronic systems to control signal flow. Switch input ports are termed poles and their output ports are termed throws. Major parameters associated with switches are:

1. The number of inputs (poles) and the number of outputs (throws);
2. Whether they are reflective type, where the off port is either open or short circuit or the nonreflective (or absorptive, terminated, or resistive) type in which the off port is set to the system impedance, which is usually 50Ω;
3. Insertion loss or on resistance;
4. Isolation or off capacitance between input and output ports.

Switch insertion loss is typically 0.5 dB or somewhat higher. Port-to-port isolation is usually at least 40 dB.

There are two main types of semiconductor device applicable to RF switches: PIN diodes or GaAs FETs.

PIN diode-based switches (PIN switches) are capable of relatively high-power operation and can be particularly cost-effective to implement. PIN diodes are described in Chapter 2. However, against PIN switches, there are the following considerations:

1. They demand a current flow to turn them on.
2. PIN switches are usually incompatible with MMIC technology and are therefore generally designed into hybrid RF circuits.

Typical bias networks for series and shunt PIN switches are shown in Figure 5.1.

GaAs FET switches have the following advantages over PIN switches:

- They implement dedicated low-current control (or enable) ports.
- Their operation is usually faster that PIN diodes (depending on the driver circuit).
- GaAs FET switches are generally compatible with RF integration (RFICs or MMICs; see Chapter 2).

GaAs FET devices are described in Chapter 2. The negative control voltages required with GaAs FET switches make the use of these devices problematic in the commonly encountered unipolar supply applications. As a result, special circuit arrangements such as capacitively floating the GaAs FETs become necessary. This approach ultimately limits the bandwidth, particularly restricting low-frequency operation. DC switching is certainly out of the question. CMOS, with its positive threshold voltage, is ideal for unipolar supplies, but the trade-off tends to be limited high-frequency operation.

PIN diode switches can be substituted for GaAs FETs but the control supply decoupling circuits add complexity.

5.2.2 Reflective and Nonreflective SPDT GaAs FET Switches

Examples of GaAs FET-based, single-pole double throw (SPDT) switches are shown in Figure 5.2. Notice there is one signal input and two signal outputs in each case.

Ideally, a FET on state is short circuit, whereas a FET off state is open circuit. The operation of the reflective GaAs FET switch is first described (the top diagram in Figure 5.2). In this circuit, there is one series FET and two shunt FETs per throw

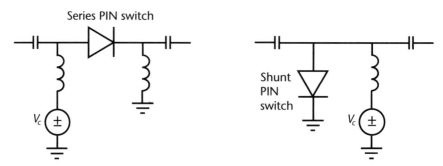

Figure 5.1 Examples of bias networks for series and shunt PIN switches [2]. (© Artech House, 2016.)

5.2 Solid State RF Switches

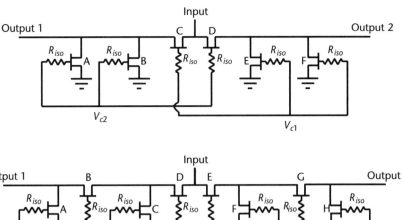
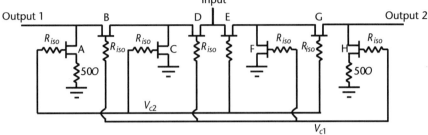

Figure 5.2 Reflective (top diagram) and nonreflective (bottom diagram) SPDT FET switch circuits [2]. (© Artech House, 2016.)

function. The control voltages are V_{c1} and V_{c2}. To enable output 1, V_{c1} is set to switch on FETs C, E, and F, while V_{c2} is set to switch off FETs A, B, and D.

As a result, the RF signal propagates across FET C and through to output 1 (leftwards in the top diagram of Figure 5.2). Meanwhile, FET D remains off, hence blocking the signal from propagating across D and through to output 2 (i.e., rightwards in the top diagram of Figure 5.2).

Because no practical device has perfect characteristics, including GaAs FETs, even when individual devices are switched off, there will be some leakage current. This is why FETs C and D are set to switch on FETs E and F, effectively shorting them to ground. The resistance of the off port in this configuration is ideally 0Ω. Further shunt elements can be added, but this approach results in a trade-off between the drive toward improved isolation versus decreased insertion loss.

The operation of the nonreflective GaAs FET switch is next described (the bottom diagram in Figure 5.2). In this circuit, there are two series FETs and two shunt FETs per throw function. Again the control voltages are V_{c1} and V_{c2}.

To enable output 2, V_{c1} is set to switch on FETs B, D, F, and H, while V_{c2} is set to switch off FETs A, C, E, and G. Otherwise, the operation is similar to that described above for the reflective case, except the off port is connected via an additional FET to 50Ω.

As mentioned above, no solid state switches can achieve perfect short or open circuits and the on and off resistances (R_{on} and R_{off}) of the FETs ultimately determine the insertion loss and the isolation characteristic of any switch. For a typical GaAs FET: a few mΩ ≤ R_{on} ≤ a few Ω and 1 kΩ ≤ R_{off} ≤ 1 MΩ.

As shown in both circuits (Figure 5.2), resistors (R_{iso}) are connected in series with every gate to prevent RF energy from leaking into the bias (i.e., gate) port. R_{iso} has to be sufficiently large to block the unwanted RF leakage and yet small enough to obviate any slowing of the switching speed. Suitable values are typically 1 to 5 kΩ.

The FET turn-on voltage is typically around 0V or a little over this value. FETs are turned off by biasing well below the pinch-off voltage, that is, generally less than − 5V.

While the foregoing discussions focus on GaAs FETs in switching mode, some switch technologies are based around CMOS technology. At least one proprietary CMOS technology claims switch operation to 60 GHz.

Microelectromechanical systems (MEMS) represent a further technological candidate for RF switches and examples are available. However, since these are fundamentally mechanical in operation there is an inherent reliability issue.

5.3 Attenuators

In order to improve matching, for example, in RF systems, various types of attenuators are required. Fundamentally attenuators reduce the signal level between processing stages in a system. This is the opposite function to amplifiers, which increase signal levels.

Three basic (passive) attenuator configurations are shown in Figure 5.3. Any one of these attenuators could be chosen to provide a specified amount of attenuation (decibels), but it is important to note that the L configuration is asymmetric, whereas configurations T and π are both symmetrical in that the input and output driving-point impedances are identical. Kingsley and Guerci [2] provide a good analysis of passive attenuator elements (pp. 180–182 of their book).

The L configuration is rarely used because it is asymmetrical and would require additional matching networks. Where A is the required attenuation (in decibels), the design equations for each resistor in the T and π configurations are

For the T configuration:

$$R_1 = Z_{out} \frac{10^{A/20} - 1}{10^{A/20} + 1} \qquad (5.1)$$

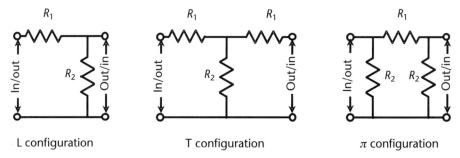

Figure 5.3 Three basic configurations for attenuator elements.

5.3 Attenuators

$$R_2 = 2Z_{out} \frac{10^{A/20}}{10^{A/20} - 1} \quad (5.2)$$

and for the π configuration:

$$R_1 = Z_{out} \frac{10^{A/20} + 1}{10^{A/20} - 1} \quad (5.3)$$

$$R_2 = \frac{Z_{out}}{2} \frac{10^{A/20} - 1}{10^{A/20}} \quad (5.4)$$

where Z_{out} is the impedance presented at the output port for each configuration.

The choice of appropriate configuration depends mainly on the realization of practical resistors. Assuming Z_{out} is 50Ω, then either of the configurations can be implemented to provide attenuation levels in the approximate range 3 to 30 dB. Resistors R_1 and R_2 both take on impractical values where the required attenuation is less than 3 dB.

Also, as the frequency increases, the parasitics associated with the resistors become increasingly significant (Chapter 3, Section 3.2.3). This feature drives the choice of resistor technology. At a high enough frequency, into the upper millimeter-wave bands, the equivalent shunt capacitor across the resistance will behave as a short circuit, hence completely ruining the function of the overall circuitry.

Step attenuators, particularly digital step attenuators (DSAs), are important because these types of circuits are frequently used in RF systems. The fundamental concept of a 4-bit step attenuator is shown in Figure 5.4.

In practice, each mechanical switch would be realized as a GaAs FET having characteristics similar to those described in Section 5.2. The overall circuit is then an electronically controlled attenuator.

The overall circuit of Figure 5.4 comprises four independent resistive attenuator elements cascaded together to form a ladder network with each attenuator having a value twice that of its predecessor (i.e., −1 dB, −2 dB, −4 dB, and −8 dB). Each attenuator network may be switched in or out of the signal path as required

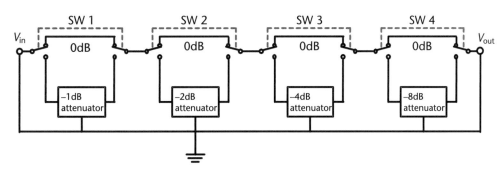

Figure 5.4 Fundamental concept of a 4-bit step attenuator.

by the associated switch producing a step adjustable attenuator circuit. Therefore, such a circuit can be switched from 0 dB to −15 dB in 1-dB steps.

As a result, the total amount of attenuation exhibited by the circuit is the sum of all four attenuator networks that are switched on. An attenuation of −7 dB would require switches SW1, SW2, and SW3 to be connected, and an attenuation of −10 dB would require switches SW2 and SW4 to be connected, and so on.

When digitally driven, such a circuit is termed a digital step attenuator (DSA) and 4-bit DSAs are commonly required in RF systems. A DSA can be connected in cascade with a variable gain amplifier (VGA) in a MMIC technology [3] in order to improve the flexibility of overall gain control.

5.4 Digital Circuits

In this section, various types of digital circuits are described. The focus is exclusively on those types of circuits that are increasingly demanded in RF systems. As stated in the introduction to this chapter, no design details are provided for digital circuits because this would be beyond the scope of this text. Reference [1] provides an excellent resource for digital electronics.

5.4.1 Selected Examples of Logic Gates

In this section, it is assumed that the reader will be familiar with basic logic gates such as AND, OR, inverters, NAND, and NOR. The simplest gate configurations embody two signal inputs, A and B, and this situation will be followed in all the circuits described here.

Because CMOS technology is fundamentally based on negative logic, NAND and NOR gates are particularly important and the basic NAND gate is now reviewed. Further technology detail regarding CMOS is considered later in this chapter. (Device details are described in Chapter 2.)

A NOT-AND operation is known as a NAND operation. It has n inputs ($n \geq 2$) and one output. The logic schematic for a two-input NAND gate is shown in Figure 5.5.

The diagram to the left of Figure 5.5 indicates a NAND gate followed by an inverter. The diagram to the right has the inverter embodied within the gate and this is the more conventional description.

All logic gates are defined by their associated truth tables. The truth table for a NAND gate is indicated in Table 5.1.

NAND gates are important in most digital systems. In particular, USB memory sticks comprise billions of interconnected and digitally addressed flash-NAND gates.

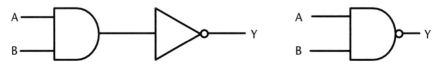

Figure 5.5 Logic schematic diagrams for a NAND gate.

5.4 Digital Circuits

Table 5.1 Truth Table for a NAND Gate

A	B	Output (Y)
0	0	1
0	1	1
1	0	1
1	1	0

The exclusive-OR (XOR or Ex-OR) gate is a special type of digital circuit. It can be used in the half adder, full adder, and subtractor subsystems of computers. It has n inputs (n ≥ 2) and one output. The logic schematic for a two-input exclusive-OR gate is shown in Figure 5.6.

The truth table for an exclusive-OR gate is indicated in Table 5.2.

One particularly important application of the exclusive-OR gate is in spread-spectrum modulation (Chapter 14).

The exclusive-NOR gate is also interesting because a 1 output only occurs when the inputs are equal. This could therefore be alternatively named the equivalence gate.

5.4.2 Digital Signal Processors

Digital signal processors (DSPs) are relatively complex types of digital circuits. These, together with all the remaining circuits described in this chapter, are generally represented within a system block or box. DSPs are therefore represented as shown in Figure 5.7.

One important application of a DSP is to generate the I and Q signal components required for *M*-ary digital modulation [4, 5], as indicated in Figure 5.8.

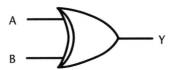

Figure 5.6 Logic schematic diagram for the exclusive-OR (or XOR, or EX-OR) gate.

Table 5.2 Truth Table for an Exclusive-OR Gate

A	B	Output (Y)
0	0	0
0	1	1
1	0	1
1	1	0

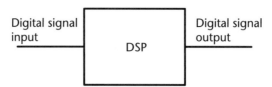

Figure 5.7 Block indicating a digital signal processor.

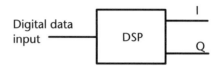

Figure 5.8 A DSP producing I and Q outputs.

The I and Q signals are then fed to individual digital-to-analog converters (DACs), which are covered in Chapter 11.

Another interesting application of a DSP is within a Weaver modulator [5, 6], which provides digital control of a single-sideband, single-carrier (SSB/SC) modulator.

Digital predistortion (DPD), often required for power amplifiers, represents a further application of what essentially comprises a DSP. The digital (DSP) portion of a DPD is shown in Figure 5.9.

This entire subsystem is essentially a DSP. The remaining parts of the DPD are presented and discussed in Chapter 10. ADCs and DACs are dealth with in detail in Chapter 11 and DPDs are further discussed in Chapter 10.

5.4.3 Electronically Programmable Read-Only Memories

In common with other relatively complex digital circuits, electronically programmable read-only memories (EPROMs)—a form of flash memory—are always indicated as a specific labeled block, as shown in Figure 5.10.

Armed with a detailed knowledge of the system requirements, the engineer programs the required algorithm into this type of chip directly from a computer. The speed is of no consequence at this stage so that standard low-frequency circuit con-

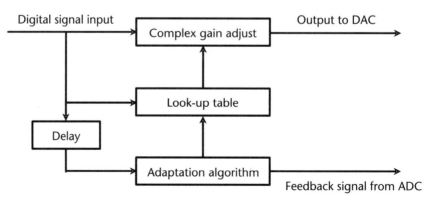

Figure 5.9 The all-digital section of a digital predistorter (DPD).

5.4 Digital Circuits

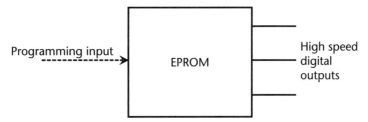

Figure 5.10 Block indicating an EPROM chip.

nections can be used. Once programmed, the EPROM is powered up in common with the entire (mainly RF) subsystem which is then in its fully operational state.

Jenkins [7] described an interesting and important application of an EPROM. In this article, focused on PLL/VCO-based frequency synthesizers, the usefulness of an EPROM is clear. The EPROM contains all the serial data to set the synthesizer parameters (more details on frequency synthesizers are provided in Chapter 12). Figure 6 of Jenkins' article [7] shows the board layout of this frequency synthesizer and the EPROM is clearly identified. The great majority of ICs having this format adopt conventional IC technology (i.e., CMOS).

5.4.4 Field-Programmable Gate Arrays

Again, in common with other relatively complex digital elements, field-programmable gate arrays (FPGAs) are always indicated as a specific labeled block, as shown in Figure 5.11.

There are several important situations within RF systems where programmed FPGAs are very useful. Examples include:

- *Look-up tables:* comprising saved files enabling the rapid retrieval of the results applying to large numbers of standard calculations;
- *Phase accumulators:* that is, clocked phase registers that are regularly updated.

Phase accumulators are key requirements, for example, within direct digital frequency synthesizers (frequency synthesizers are covered in Chapter 12).

FPGAs generally support common mathematical functions (EXP, LN, LOG, SQRT, SIN and logic functions including NAND, NOR, and XOR). Historically, application-specific integrated circuits (ASICs) have been implemented for the above functionality, but this approach is decreasing in favor of FPGAs.

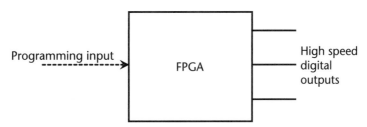

Figure 5.11 Block indicating an FPGA chip.

5.4.5 Provision for Built-In Test and Related Requirements

Digital technology is particularly well suited for implementing various built-in test (and related) facilities. The options are built-in test (BIT), built-in self-test (BIST), and built-in test and repair (BITR).

Any of these options will require sampled signals taken from strategic locations around the RF circuits. The sampled signals are usually obtained by means of dedicated couplers that are very loosely coupled to avoid any significant interruption of the signal. For example, such couplers can be prelocated at the input and also the output of amplifiers. The output of each coupler connects to the input of an ADC, the output of which is fed to an FPGA chip. This BIT (or BIST) subsystem will detect whether the amplifier gain is degraded and will send an error message to the system manager's dashboard. At this stage, a backup RF system could automatically be switched on so the wireless network would remain operational.

A built-in test and repair (BITR) subsystem embodies most of the above elements, but goes an important step further in terms of actually varying an RF circuit parameter to overcome a detected degradation in performance. The BITR subsystem could, for example, adjust a bias voltage to compensate for an unwanted decrease or increase.

5.4.6 Technology Utilized for Digital Circuit Elements

As indicated above, in relation to EPROM chips, the great majority of digital ICs adopt conventional IC technology (i.e., CMOS). Minimum feature size largely dictates maximum possible (CMOS) switching speed and it is possible to obtain ICs with feature size around 7 nm of less. FinFET technology, in which the MOS individual transistors possess 3-D topology enabling faster operation, have also entered the market. Clocking at frequencies up to about 10 GHz is commonly encountered on-chip or up to around 4 GHz on PCBs.

The switching times associated with the individual transistors are one fundamental factor determining maximum operating frequencies. As these switching times decrease, typically below 1 ns, then higher-frequency operation becomes possible. Towards the end of Section 5.2, it is stated that at least one proprietary CMOS technology claims switch operation to 60 GHz. At this frequency each cycle occupies just under 12 ps, which means the maximum switching times of each transistor in these circuits has to be in the region of 1 to 2 ps. In this technology, the individual transistors have extremely fast switching capabilities.

Beyond the transistors, the interconnects can often be the limiting factor in terms of circuit speed. A deep and detailed appraisal of interconnects for digital systems is provided in Chapter 21 of [8].

It is also very important to appreciate that CMOS technology is extendable to realizing amplifiers, oscillators, and many other analog circuit functions, including operation into millimeter-wave bands.

By the mid-twenty-first century, radically new digital technological approaches will likely have gained ground, for example, quantum-based devices and circuits.

References

[1] Lombardi, F., and J. Huang, *Design and Test of Digital Circuits by Quantum-DOT Cellular Automata*, Norwood, MA: Artech House, 2007.

[2] Kingsley, N., and J. R. Guerci, *Radar RF Circuit Design*, Norwood, MA: Artech House, 2016.

[3] "Variable Gain Amplifiers for Wireless Infrastructure, Product Description by Integrated Device Technology," *Microwave Journal*, December 2016, pp. 38 and 40.

[4] DeMartino, C., "The Differences Between Transmitter Types, Part 2," *Microwaves & RF*, April 2017, pp. 69–70 and 83.

[5] Popovic, Z., "Amping Up the PA for 5G," *IEEE Microwave Magazine*, May 2017, pp. 137–149.

[6] www.arrl.org/files/file/QEX_Next_Issue/2014/Nov-Dec_2014.

[7] Jenkins, R., "Compare Phase-Locking Methods of RF Sources," *Microwaves & RF*, April 2017, pp. 77–80.

[8] Edwards, T., and M. Steer, *Foundations for Microstrip Circuit Design*, 4th ed., New York: John Wiley & Sons, 2016.

CHAPTER 6

Radio-Frequency Filters

6.1 Introduction

This chapter focuses mainly on fixed-frequency filter requirements and designs. However, in a later section, the state of the tunable filter art is critically discussed. Substantial material has been based on the M.Sc. course notes that I prepared and taught at Hull University, United Kingdom [1].

6.2 Review of Basic Concepts and Fundamentals

Thinking for a moment beyond electronics, the most basic concept of a filter is an arrangement that enables one to input perhaps a load of material that comprises some that is ultimately required, while leaving behind waste material that may be discarded. RF filters enable those frequencies actually required for subsequent processing in the system to be transferred to the next stage, while largely discarding any frequencies that are not desired.

The four basic classes of frequency filters are:

- BPFs;
- Bandstop filters (BSF);
- Highpass filters (HPF);
- Lowpass filters (LPF).

Figure 6.1 shows the standard circuit symbols representing each of these basic classes of filter.

Free-wave symbols represent signal pass frequency bands, whereas struck-through waves represent bandstop characteristics. In each case, imagine a vertically disposed axis indicating increasing frequency from bottom to top in each symbol box. For the BPF, there is a specific passband with lower and upper stopbands. In the case of the BSF, there is just one stopband. For the LPF, all frequencies are passed from DC up some specified frequency limit. HPFs are essentially the mirror-image of this in that a range of lower frequencies are all stopped and only a specified higher band of frequencies are passed.

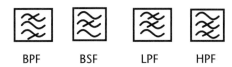

Figure 6.1 The four basic classes of frequency filter.

These symbols are extensively adopted throughout this book. In the RF context, BPFs, BSFs, and LPFs are the most important classes of frequency filter.

The concept of ideal filter characteristics represents an important baseline, as follows:

- Zero loss in the passband;
- Precise start and stop frequencies;
- Infinite attenuation immediately available at the skirts;
- Infinite attenuation maintained throughout the stopbands;
- No spurious or harmonic responses;
- Perfectly linear frequency response through the passband.

These concepts are illustrated (for a BPF) in Figure 6.2.

It is impossible to achieve rapid (immediate) changes in insertion loss as frequency is altered. In the time domain, clearly everything actually takes time (i.e., a truly immediate change in some variable [literally taking zero time] is nonsense). Correspondingly, in the frequency domain, a precisely immediate change in some variable (literally taking zero frequency) is also nonsense, let alone having the insertion loss suddenly becoming infinite. Neither is it possible to have the ideal situation where there is zero insertion loss throughout the passband of any filter.

Instead, in practice, an RF BPF spectral characteristic will take on a shape more like that indicated in Figure 6.3.

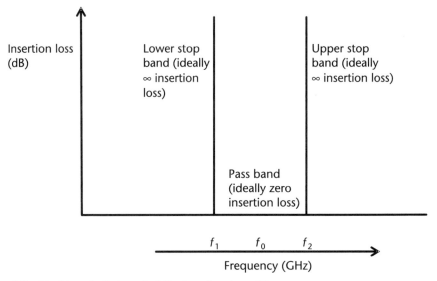

Figure 6.2 Ideal (spectral) characteristics for a bandpass filter.

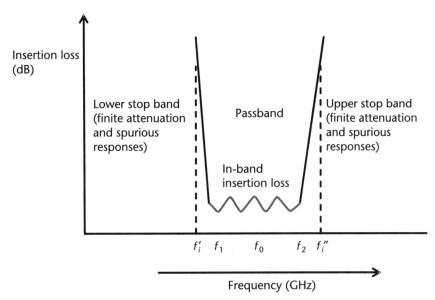

Figure 6.3 Typical (practical) frequency characteristic of a BPF (center frequency f_0).

The characteristic shown in Figure 6.3 illustrates the following unavoidable facts:

1. There will always be finite insertion losses (decibels) at all frequencies.
2. There can never be sudden changes in these insertion losses.

It is also worth bearing in mind that the stop band attenuations might typically be in the region of several tens of decibels, while the in-band insertion loss might amount to anything from a fraction of a decibel through to several decibels. This in-band (or passband) insertion loss usually comprises two elements: loss due to attenuation in the structure of the filter, and a ripple element that can be caused by the choice of filter design.

It is also apparent from Figure 6.3 that there are two distinct bandwidths: the passband ($f_2 - f_1$) and the attenuation bandwidth ($f_i'' - f_i'$).

In the foregoing definitions, attenuation (or insertion loss) symmetry is assumed (i.e., f_i' and f_i'' are disposed symmetrically about f_0).

In practice, close dimensional control (especially the effects of tolerances) is vital for the design success of any type of filter. Losses lead to potential problems with all forms of microwave filters.

6.3 Technology Options

Basic LC networks (e.g., ladder networks) can be used for the design and fabrication of many types of filters. At relatively low frequencies (through some hundreds of megahertz), such LC networks are practical propositions. The L and C components are designed in accordance with the details provided in Chapter 3.

At microwave frequencies, inherent losses in the L and C elements, leading to low Q-factors, render this approach ineffective and therefore distributed, transmission line structures are preferred. Contrastingly, for realizations at millimeter-wave frequencies (60 GHz and above), networks implementing lumped L and C have been proven to be successful and this approach is considered in more detail later in this chapter.

How to immediately recognize (with a lumped-component design) whether a filter is basically lowpass of highpass: if a series inductor is involved (with no series capacitor), then DC can get through and this must be a lowpass filter. Otherwise, the design must be for a highpass filter.

The next section looks at the use of microwave line sections to realize LPFs.

6.4 LPFs Formed with Cascaded Microstrips

The basic concept here is to employ a cascade of microstrip transmission sections, simulating series L, shunt C, types of elements, as indicated in Figure 6.4.

In a general note regarding design procedure, all filter design procedures that will be covered here begin with selecting the appropriate lowpass (low frequency) normalized filter prototype characteristics.

This is a fundamental approach that follows the detailed guidelines laid down, in particular, by the classic work of Matthaei et al. [2]. In particular this approach yields the basic prototype data.

The next step is to follow the insertion loss synthesis procedure.

For lowpass filter design, the procedure is:

1. Select the appropriate prototype for the desired frequency response characteristics. This yields normalized design values for the lowpass network. Ls are in henries, and Cs are in farads. The resulting network would give the final decibel values required, but in hertz units and for 1-Ω characteristic impedance.
2. Transform for the desired microwave frequency band and 50Ω. Ls and Cs decrease directly by the frequency factor and 50-Ω terminations now apply, meaning a multiplication 50 times the initial 1-Ω value. This yields a lumped network that (ideally) would provide the response.
3. Realize the results of step 2 implemented using suitable microwave technology (e.g., cascaded microstrips for LPF).
4. For an LPF, each cascaded microstrip element must be $< \lambda_g/4$ in length (any longer than this results in the element departing from quasi-lumped-equivalent behavior).

It is always important to recollect the fact that the concern is with realizing, and often transforming from, sections of microstrip or other circuit elements having

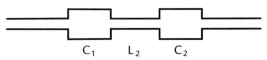

Figure 6.4 Cascaded sections of microstrip lines effectively simulating two capacitors and one inductor (i.e., a π – network).

6.4 LPFs Formed with Cascaded Microstrips

physical length l and characteristic impedance Z_0. In practice, all the elements will have attenuation (power losses) as well as reactive properties. However, it is possible to effectively design by first assuming negligible loss and then introducing the loss mechanisms once each filter design is initially completed.

For assumed loss-free lines then an equivalent $\pi - section$ reduces as shown in Figure 1.9 (LPFs).

Because DC can be transmitted, it should immediately be apparent this is already a simple, basic, LPF structure.

Figure 1.9(c) shows the layout in microstrip format and the step discontinuities between low and high impedances is itself capacitive, contributing to the capacitance desired in the LPF design.

In general, high-impedance lines behave as series inductances L and low-impedance lines behave as shunt capacitances C.

For the series inductive section of microstrip line, equate L to the fundamental expression for the transmission line inductance and rearrange for ℓ to yield:

$$\ell = \frac{\lambda_g}{2\pi}\sin^{-1}\left(\frac{2\pi fL}{Z_0}\right) \qquad (6.1)$$

Note that in the final LPF structure ℓ is termed ℓ_L.

When the physical length is less than one-eighth wavelength, using the small-angle approximation gives us the simple result:

$$\ell \cong \frac{f\lambda_g L}{Z_0} \qquad (6.2)$$

For the series capacitive section of microstrip line equate capacitance C to the transmission line capacitance and rearrange for ℓ to yield:

$$\ell = \frac{\lambda_g}{2\pi}\sin^{-1}(2\pi fCZ_0) \qquad (6.3)$$

or, again usefully, especially since these line lengths are almost always very short electrically and physically:

$$\ell \cong f\lambda_g Z_0 C \qquad (6.4)$$

This analysis leads to the cascaded microstrip layout shown in Figure 6.5.

In the final LPF structural diagram of Figure 6.5 the capacitive length of line is termed ℓ_C. It is also assumed that $C_1 = C_2$ and the signal inputs and outputs are both terminated in 50Ω.

All microstrip widths are known functions of their individual characteristic impedances (Chapter 3). It is also important to allow for end-step discontinuities because these add extra capacitance.

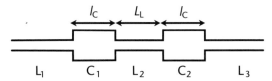

Figure 6.5 Microstrip layout of a basic LPF indicating the physical lengths and the corresponding lumped components.

An entirely different approach to LPF design leads to cascades of microstrip stub elements, distributed along the main microstrip line. The basic type of resulting structure is shown in the lower diagram of Figure 6.6.

The low-high-low structure (top of Figure 6.6) shows that the capacitive microstrip lines can be unequal where necessary or desirable. The multiple-stub approach may provide improved matching capability is some instances.

Radial stubs are yet another option for implementing as filter elements. The design approach for radial stubs is provided in Chapter 3.

6.5 Microwave BPFs

The most fundamental form of BPF is the basic LC resonator. This can comprise series or parallel LC, as shown in Figure 6.7.

Each of these circuits will resonate at a particular frequency, passing or rejecting a signal, depending on whether the arrangement is series or parallel. This represents a basic and rather crude form of BPF.

However, the introduction of a cascade of such resonant circuits (usually the parallel form) can lead to a prescribed BPF design. The most usual such cascade comprises parallel-coupled half-wave microstrip resonators. It is necessary to begin with the prototype lumped-element BPF structure [2] shown in Figure 6.8.

However, there is a serious practical difficulty associated with this configuration. This difficulty arises from the fact that each half-wave microstrip resonator

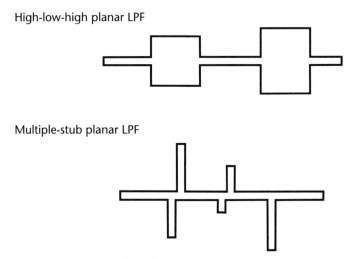

Figure 6.6 Low-high-low (top) and alternate stub LPF layouts.

6.5 Microwave BPFs

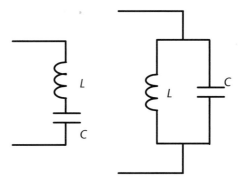

Figure 6.7 Series and parallel resonant LC circuits (parasitic resistances are omitted).

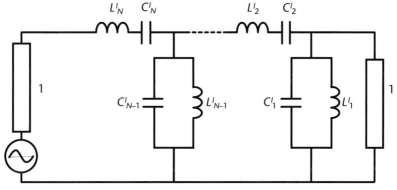

Figure 6.8 Lumped-element prototype BPF.

can only exhibit one or the other resonance behaviors: either series-like or parallel-like, but not both in the same system. The design approach must accommodate this fact.

With microstrip and other TEM-like coupled structures, the following conditions are required:

- Maximum coupling using $\lambda_g/4$ (or some odd multiple thereof) coupled regions;
- Each resonator must be $\lambda_g/2$ in length (or any multiple thereof).

Figure 6.9 is a plan view showing two coupled microstrip resonators.

This type of structure gives rise to two distinct field modes in the coupled region: odd and even (described in Chapter 4), which result from the two extremes of field distributions between the strips. The even mode has the associated characteristic impedance Z_{0e}, while the odd mode has its (always different) characteristic impedance Z_{0o}.

Knowing these two impedances, the widths (w) and spacings (s) between the coupled lines can be calculated (a process similar to that described in Chapter 3, for single microstrip lines). An iterative process (using an EDA program) enables the convergence toward the final design.

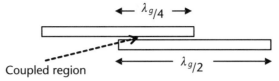

Figure 6.9 Plan view indicating two coupled microstrip resonators.

A basic parallel-coupled microstrip BPF will therefore take on the general layout shown in Figure 6.10.

In this structure $\ell_1, ..., \ell_4 \cong \lambda_g/4$.

Note that each parallel-coupled section, ℓ_1, w_1, s_1 and so on, constitutes one section.

Detailed design is based on four main steps:

1. Determine the one-type resonator network, to realize the specification, from the original prototype.
2. From the network parameters, evaluate the even- and odd-ordered characteristic impedances Z_{0e} and Z_{0o} applicable to the parallel-coupled microstrip.
3. Relate the values of Z_{0e} and Z_{0o} to microstrip widths and separations (w, s) (detailed microstrip design).
4. Calculate the whole resonator length 2ℓ, which is slightly less than $\lambda_g/2$ because of the semi-open end effects, and therefore of the coupled-section length ℓ, which is slightly less than $\lambda_g/4$ for the end-effect reason again.

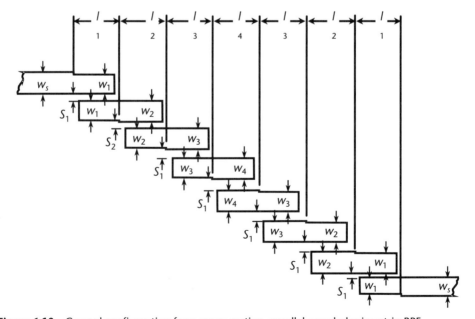

Figure 6.10 General configuration for a seven-section, parallel-coupled microstrip BPF.

In the above, λ_g is the mid-band and average microstrip wavelength. Allowance must be made for the semi-open-circuit microstrip end-effects, which exist for all elements in this circuit.

Defining δ as the fractional bandwidth:

$$\delta = \frac{f_2 - f_1}{f_0} \tag{6.5}$$

The frequency transformation from the lowpass prototype filter to the bandpass microwave filter is then:

$$\frac{f_i'}{f_c'} = \frac{2}{\delta}\left(\frac{f_i - f_0}{f_0}\right) \tag{6.6}$$

Noting that f_c' is the prototype cutoff frequency (it equals: $1/2\pi$). So this can always be substituted early in the design process.

However, f_c' has to be defined in the initial filter specification (it defines the attenuation/frequency gradient beyond cutoff). In practice the parameter ($\left|\frac{f_i'}{f_c'} - 1\right|$ is used in determining the selection of basic prototype filter characteristics that will best fit that desired for the final microwave filter).

To proceed with the microstrip design, the remaining requirements are to determine the odd- and even-mode coupled-line impedances Z_{0e} and Z_{0o}, which are given by equations interrelating the system characteristic impedance Z_0 (usually 50Ω, that is, that of the lines feeding the filter) and the coupling parameters.

This leads to the values (per section) of Z_{0e} and Z_{0o} and hence the microstrip widths and spacings throughout the structure.

NI AWR provide an example of a parallel-coupled (or edge-coupled) microstrip BPF to be realized on a Rogers TMM 10 substrate. The filter is designed to operate over the 12–18-GHz band.

The input and output sections of the layout for this BPF are shown in Figure 6.11. The output characteristics of this filter are summarized: insertion loss ≤–0.6 dB across the passband, and rejection characteristics (skirts), 10 dB at 12.2 and 13 GHz. Otherwise, 16 dB is the worst case at 14.2 GHz.

In practice, most microstrip or related planar BPFs have their topologies rotated for spatial convenience (on chip or substrate) as shown in Figure 1.10.

In the description of Figure 1.10, it is interesting to note the alternative terminology (i.e., specifying the structure in terms of the number of resonators rather than the number of sections). The angled layout is very often used because it saves surface space on the substrate.

U-shaped (or hairpin), one-wavelength-long resonators can be implemented for reduced radiation loss by optimizing the end-U separations. However, this takes more chip area (real estate) and is not advised for MMIC realization.

Parallel-coupled (or edge-coupled) microstrip BPF configurations can yield passband widths up to at least 10% (e.g., 2.8 GHz on 28-GHz center frequency).

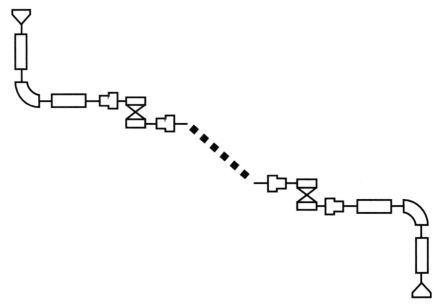

Figure 6.11 The 12–18-GHz edge-coupled filter rotated 45°. (Courtesy of NI AWR.)

6.6 Suspended Substrate Stripline Filters

Suspended substrate stripline (SSS) is a particularly low-loss transmission line technology. Unlike microstrip (which is comparatively lossy), SSS or S^3 has the center strip (or strips) supported on a thin low-loss dielectric substrate. This structure does not require a ground plane attached to the back side of the substrate but instead it is surrounded by a metal box.

Two parallel-coupled SSS are shown in the cross-section in Figure 6.12.

The metallic parts can be manufactured from copper or aluminum. The substrate usually comprises a stiffened polymer material that is sometimes glass-fiber reinforced. This substrate, fixed mechanically at both ends, is usually much thinner than shown in Figure 6.12. Ideally, it would comprise an infinitely thin sheet (but this is hard to achieve in practice).

The metal box is grounded and mode suppression screws and other elements are installed inside this box to suppress unwanted field distributions such as rectangular waveguide modes.

Using the S^3 structure, BPFs can be designed according to the methodology described above for the microstrip case, although with S^3 the design is actually more straightforward because the even-mode and odd mode field distributions are almost identical.

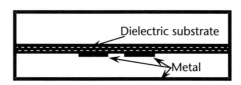

Figure 6.12 Cross-section of suspended-substrate stripline.

However, it is important to observe that filters (mainly BPFs) manufactured using this structure cannot usually be integrated into an MIC or an MMIC. These types of filters are most often implemented as (or within) a stand-alone module.

6.7 Inline Microstrip Filter Structures

In this section, special structures are considered, built within the width of a microstrip line. In particular, spur-line bandstop (or notch) filters are examined. The basic resonant structure for this is indicated in Figure 6.13.

This approach to the design and manufacture of BSFs ensures the following advantages:

- The structure is almost completely nondispersive, that is, the guided wavelength λ_g changes practically linearly with frequency (see Chapter 3).
- The structure is readily manufactured within the width of the microstrip line.
- Being largely constrained within the microstrip, losses due to radiation are relatively low.

The central structure of length a is equivalent to an open-circuit shunt stub of characteristic impedance Z_1 followed by a continuous section of microstrip having a characteristic impedance Z_{12}. In common with the BPF design outlined above so also here the analysis is not presented in detail. However, some important results are:

The length a (see Figure 6.13) is electrically $\lambda_g/4$ at mid-band (i.e., f_0) although the gap equivalent length ℓ_{eg} will reduce the physical value of a such that

$$a = \frac{3 \times 10^8}{4f_0\sqrt{\varepsilon_{eff0}}} - \ell_{eg} \text{ in m} \quad (6.7)$$

where ε_{eff0} is the odd-mode effective microstrip permittivity at the center frequency and ℓ_{eg} is the equivalent gap end-effect extension.

The value of the series gap separation b has to be determined empirically and a value of approximately 50 μm appears to be suitable for designs on alumina, and similar substrates [3]. Once b is chosen, the end-effect length may be found by using

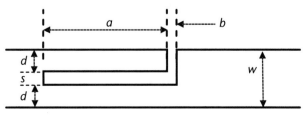

Figure 6.13 Basic structure of the spur-line BSF.

$$\ell_{eg} = C_{odd}v_{po}Z_{0o} \tag{6.8}$$

C_{odd} is calculated from expressions and data given by Benedek and Silvester [4] and Bates [3] gives results that indicate that a very good agreement between calculation and experiment can be achieved for gap separations of about 100 μm or less.

The BSF performance is particularly good when two of these spurline structures are arranged back to back in cascade, with a separation slightly less than a quarter-wavelength. For example, where a = 3.2 mm, the distance separating the structures should be about 3.0 mm [3]. With such a filter, a rejection (insertion loss) > 35 dB was achieved over the band: 9.3 $\leq f \leq$ 9.7 GHz.

Several variations of the structure are conceivable. A double-sided arrangement (i.e., two spur-line structures side by side across the microstrip width) has been manufactured for a band centered on 30.5 GHz. The rejection exceeded 29 dB over a 2-GHz band [3].

Such compact and effective BSFs are very important in several applications, including microwave and millimeter-wave mixers (Chapter 13).

This general concept, leading to unusual and very compact filters, must be capable of extension into many new designs. Microstrip lends itself to entirely new structural shapes to perform circuit/system functions, which were previously realized using triplate (stripline), or coaxial, or waveguide technology.

6.8 Filters Using Defected Ground Plane Technology

For several decades, the main focus of integrated microwave engineering remained almost exclusively on the microstrip (top) lines. Eventually, it was appreciated that performance advantages could be obtained by refocusing on the ground plane and strategic changes introduced into this important plane became known as defected ground plane (DGP) or alternatively defected ground structure (DGS) technology.

The basic concept of DGP (or DGS) is indicated in Figure 6.14.

Figure 6.14(a) shows the top microstrip line of width w and small central gap g. A slot of width a is also shown, cut into the ground plane beneath the microstrip line. This slot protrudes distance b away from the edge of the microstrip on both sides symmetrically. This represents the most basic structure of a DGP-enhanced microstrip and the frequency characteristics are indicated in Figure 6.14(b). It can be seen that even this relatively simple structure begins to display filter-like characteristics.

Many variations have been reported on the cross-DGP configuration such that useful filters have been realized (mainly LPF but also some BPF).

Work reported by Weng et al. in 2008 [5] and Boutejdar et al. in 2017 [6] provide extensive updates on DGP (DGS) technology.

6.9 Dielectric Resonators and Filters Implementing Them

Dielectric resonators (DRs) are important in filters and also for some types of microwave oscillators (DROs, covered in Chapter 12).

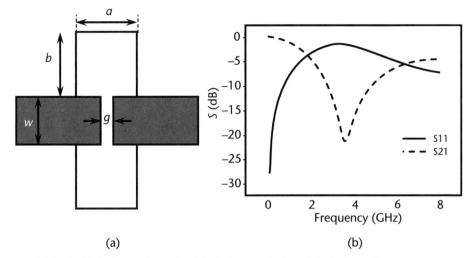

Figure 6.14 (a, b) Basic structure of a defected ground-plane (DGP) element.

Any cylindrically shaped dielectric structure can be caused to resonate at some specific frequencies and these depend upon both the permittivity and the physical dimensions. The basic concept is illustrated in Figure 6.15.

The resonant frequency is dependent on the radius a, height d, permittivity ε_r, free-space velocity of light c, the number of half-wavelengths supportable in the vertical direction, and a Bessel function parameter known as $\chi_{m,n}$. Further details concerning the analysis will not be followed through here, but the important feature to consider is that this simple structure is indeed resonant.

In practice fringing fields, microstrip (or other) line coupling and the presence of the substrate dielectric render the analysis complex and more sophisticated computer-based procedures are necessary to evaluate the applicable parameters. A practical DR is usually cylindrical in form and is mounted in a metallic enclosure having height H. When the thickness of the DR is t, then the ratio H/t should be approximately 3 for optimum operation.

Liang [7] provides data on a range of nine commercially available DRs. Relative permittivities ranged from 21 to 47 and resonant frequencies ranged from 5 to 10 GHz. The figure of merit (FOM) Qf applying to the resonators occupied a tenfold range, coming in as low as 30,000 but also as high as 300,000. This means that the unloaded Q-factors can be anything between 6,000 and 30,000, depending on losses (mainly originating from impurities) in the dielectrics.

A five-part dielectric resonator BPF is shown coupled to feeding microstrip lines in Figure 6.16.

Figure 6.15 A cylindrical dielectric resonator.

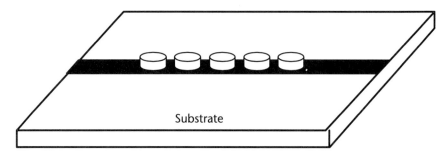

Figure 6.16 A BPF realized as a cascade of coupled dielectric resonators.

Figure 6.16 shows 5 times coupled DRs mounted on a substrate with input and output microstrip feed lines. All the dense black features are microstrip lines and this diagram is not to scale; it only provides the concept of the layout.

However, loaded Q–factors depend strongly upon the degree of coupling to the microstrip line, (i.e., the proximity of the two elements). When the coupling distance is less than 1 mm, the loaded –factor is usually below 500 and this has significant consequences in filter and oscillator design.

The BPF design approach described above (for parallel-coupled microstrip implementation) is also followed initially for DR filters. The optimum coupling for each section (adjacent resonators) is determined using this same basic insertion loss design procedure.

To increase the coupling while avoiding excessive proximity of the microstrip lines, curved coupling sections are used. Each DR is also optimally coupled and, as with all filters, a major design aim is to identify appropriate DRs and to establish the degrees of coupling applicable to each section.

There are generally difficulties associated with in-band insertion loss (ripple) mainly because of the relatively low loaded-Q-factors of the DRs.

6.10 SIW-Based BPFs

The fundamentals of substrate-integrated waveguide (SIW) are provided in Section 3.5 of Chapter 3.

BPFs have been developed implementing SIW, in particular miniaturized BPFs using quarter SIW resonator-based designs incorporating an elliptic defected structure [8]. A layout of this type of BPF is shown in Figure 6.17.

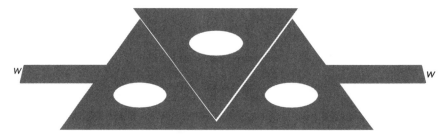

Figure 6.17 Cross-coupled cascaded triplet SIW BPF [8].

The total area occupied by this filter is 20 mm by 50 mm (just under 1 inch by 2 inches) and the 0.508-mm-thick substrate has a permittivity of 2.2.

The center frequency is 3.6 GHz and the 3-dB bandwidth is 8.5%. The in-band insertion loss is 1.2 dB, which is somewhat high; [8] is recommended for further data.

6.11 Millimeter-Wave BPFs

In general, as frequency increases, so losses also increase and Q-factors decrease and this trend poses serious challenges for all types of frequency filters. In particular, in-band insertion loss tends to increase, unwanted signal rejection is compromised, and skirt selectivity is degraded.

Unless there is a radical breakthrough in ADC capability, millimeter-wave communications technology, effectively starting around 27 GHz, will likely remain rooted in traditional RF receiver architecture (Figure 6.15). This will mean that fixed RF BPF specifications will continue to apply. Even so, millimeter-wave filter design (notably millimeter-wave BPFs) remains a serious challenge and it is very hard to achieve <1-dB passband insertion loss, for example.

Chaturvedi et al. [9] provides a detailed survey of passive millimeter-wave bandpass filters, covering many semiconductor-based, MMIC-compatible approaches. As a specific example, Chaturvedi et al. [9] presents the CMOS-based BPF topology originally due to Lu et al. [10] and this is indicated in Figure 6.20.

Miniature MIM capacitors and microstrip line-based inductors are implemented to form the circuit of Figure 6.18. The following results apply to this BPF: center frequency 60 GHz, 3-dB bandwidth 18.28%, in-band insertion loss 2.55 dB, and chip area occupied 0.085 mm^2.

Zhang et al. [11] published an extensive, in-depth appraisal of the three-dimensional (3-D) printed filters (and other related devices) scenario.

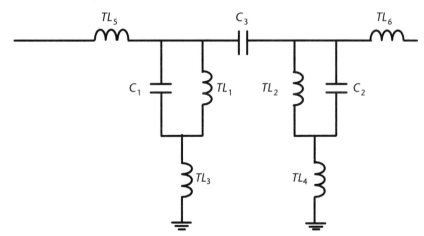

Figure 6.18 Topology of a miniaturized millimeter-wave BPF incorporating an internal π–circuit and realized using standard 0.13-μm CMOS technology.

6.12 Tunable BPFs

This important topic continues to receive close attention from research and development groups. Relating back to Figure 1.8, it can be seen that there is an increasingly strong demand for tunable BPFs capable of operating at both microwave and millimeter-wave frequencies.

Broadband-tunable filters have been commercially available for many decades. These are based on the somewhat exotic ferrite material known as YIG (yttrium-iron-garnet), which comes in the form of tiny spheres that require controlled magnetization to tune them. This requirement means the realization of these devices inevitably means modules having substantial physical sizes. YIG-tuned filters (also oscillators) are therefore incompatible with integrated assemblies and certainly not compatible with MMIC technology. These types of devices remain useful in instrumentation and military (electronic warfare [EW]) applications. They are also relatively expensive, attracting unit prices upwards of a few hundred U.S. dollars.

Therefore, considerable effort continues to be expended towards realizing tunable filters that would be compatible with integrated RF electronics. Examples include the use of barium strontium titanate (BST) MIM structures integrated onto lithium niobate to form electrically variable capacitors. This results in a form of nonsemiconductor varactor that can tune a surface-acoustic wave (SAW)-based BPF to operate over 999 and 1,018 MHz. Using the same technology but pushing the frequency up to the 6.2–7.5-GHz band results in unacceptably high passband losses ranging from 6.3 to 8.5 dB.

It has also been shown that similar technology can be used to tune an ultrawideband (UWB) BPF over the 8.327–13.942-GHz band.

The July/August 2014 issue of *IEEE Microwave Magazine* [12] "Flexible Variety of Filters: Recent Design Trends" provided an excellent resource for leading articles profiling in detail just where filter research and design is heading.

References

[1] Extensive material from the M.Sc. Course on Radio Frequency Systems notes prepared for and taught by the author, at Hull University U.K. early 2000s.

[2] Matthaei, G., L. Young, and E. Jones, *Microwave Filters, Impedance-Matching Networks and Coupling Structures*, New York: McGraw-Hill, 1965, reprinted in 1980, Dedham, MA: Artech House.

[3] Bates, R., "Design of Microstrip Spur-Line Band-Stop Filters," *IEE Journal on Microwaves, Optics and Acoustics*, Vol. 1, No. 6, November 1977, pp. 39–41.

[4] Benedek, P., and P. Sylvester, "Equivalent Capacitances for Microstrip Gaps and Steps," *IEEE Trans. on Microwave Theory and Techniques*, Vol. 20, No. 11, November 1972, pp. 729–733.

[5] Weng, L. H., et al., "An Overview on Defected Ground Plane Structure," *Progress in Electromagnetics Research B*, Vol. 7, 2008, pp. 173–189.

[6] Boutejdar, A., and S. Dosse Bennani, "Design High-Frequency Filters with Cross DGS Circuit Elements," *Microwaves & RF*, May 2017, pp. 76–85 and 131.

[7] Liang, E. C., "Characterization and Modeling of High Q Dielectric Resonators," *Microwave Journal*, November 2016, pp. 68–86.

[8] Zhang, S., et al., "Miniaturized Bandpass Filter Using Quarter SIW Resonator with Elliptic Defected Structure," *Microwave Journal*, March 2017, pp. 74–80.

[9] Chaturvedi, S., M. Bozanic, and S. Sinha, "Millimeter Wave Passive Bandpass Filters," *Microwave Journal*, January 2017, pp. 98–106.

[10] Lu, M. C., et al., "Miniature 60 GHz-Band Bandpass Filter with 2.55 dB Insertion Loss Using Standard 0.13 μm CMOS Technology," *Microwave and Optical Technology Letters*, Vol. 51, No. 7, April 2009, pp. 1632–1635.

[11] Zhang, B., et al., "Review of 3D Printed Millimeter-Wave and Terahertz Passive Devices," *International Journal of Antennas and Propagation*, Volume 2017 (2017), Article ID 1297931, https://doi.org/10.1155/2017/1297931.

[12] "Flexible Variety of Filters: Recent Design Trends (All Papers)," *IEEE Microwave Magazine*, Vol. 15, No. 5, July/August 2014.

CHAPTER 7
Antennas

7.1 Introduction

RF technology underlying communications systems would be almost useless unless signals could be radiated from a transmitting antenna and then received by a receiving antenna (sometimes the same device). The obvious exception is cabled systems, but the great majority of RF networks utilize access to air (or space) links, and this means antennas.

It is important to bear in mind that all forms of electronics radiate electromagnetic energy. In cases in which circuitry, the main subject throughout this book, is being considered, a major design aim is to ensure that radiation is maintained below acceptable levels. Antennas come in where the opposite is demanded (i.e., the structure is designed to efficiently radiate electromagnetic energy). Any antenna is also electrically reciprocal (i.e., it can transmit as well as receive RF signals).

A reasonably detailed understanding of antennas is essential for almost any RF engineer and the purpose of this chapter is to support this need as far as reasonably possible in a single chapter. Many earlier texts provide an extensive coverage of this subject, notably Balanis [1] and Pozar [2]. In particular, Pozar includes a deep and detailed treatment.

The following key definitions apply to antennas:

- Antenna or radiation pattern: a two-dimensional (2-D) or three-dimensional (3-D) plot of field strength along the radiation pattern out into space beyond the antenna's aperture;
- Antenna noise temperature: the combined effect of external noise (notably, sky noise) and internal antenna noise sources;
- Aperture: the effective electrical area at the front-end of an antenna;
- Azimuth (AZ): the direction of a reference point looking along the XY-plane (defined as from 0° to 360° or −180° to +180°);
- Bandwidth: the range of frequencies over which an antenna will operate to specification;
- Beam (or lobe): the volume of radiated energy directed away from the antenna;

- Beamwidth: the width of the main beam (in degrees) where the antenna gain is at N dB of the maximum (usually 3 dB);
- Boresight: the direction is which the antenna is pointing;
- Boresight error: the difference between the boresight direction and the direction of maximum radiated intensity;
- Directivity: the ratio of the field strength at boresight (typically) compared with the field strength averaged over all directions;
- Elevation (EL): the direction of a reference point looking up and down or along the Z-axis (expressed from –90° to +90°);
- Gain (G): the ratio of the field strength at boresight (typically) compared with the field strength of an isotropic antenna;
- Input impedance: ideally this should be identical to the characteristic impedance of the feed system (usually 50Ω) (in practice, a matching network is often required);
- Isotropic antenna: an antenna that radiates with precisely equal field strength in all directions (unachievable in practice but an important comparative baseline);
- Omnidirectional antenna: an antenna that radiates with precisely equal field strength in all azimuthal directions (the result appears as a circle on a spatial plot of the antenna radiation);
- Peak sidelobe ratio: ratio of intensities of the highest (most prominent) sidelobe to that of the main beam;
- Radiator: the portion of an antenna that radiates, that is, propagates electromagnetic energy (same as aperture);
- Range: the distance from the antenna to some prescribed reference point;
- Reflector: a special part of many antennas that serves to focus the radiated energy in a desired direction (good examples include most multi-element arrays such as Yagi and parabolic reflectors);
- Return loss: the ratio of the power reflected by an antenna to the power fed into it;
- Sidelobe: a pattern of radiation in an undesired direction.

Several of the above definitions are revisited in detail in the following sections. The coverage of this chapter ranges from antenna fundamentals through to phased-arrays including beamforming and patch arrays.

7.2 Antenna Fundamentals

The signal to be transmitted by any transmitting antenna generally originates from an RF power amplifier (RFPA), which is considered in Chapter 10. Between this RFPA and the antenna, there are often fairly complex feed networks but the arrangement can be simply considered as shown in Figure 7.1.

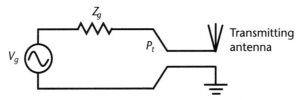

Figure 7.1 Fundamental outline of an antenna in transmitting mode.

In Figure 7.1 the source generator has RMS voltage V_g and impedance Z_g. The transmitting power is P_t and the transmitting antenna is shown as the general symbol (i.e., three elements) spreading out. Although Figure 7.1 (and many examples in this chapter) refers to transmitting antennas, most of the definitions and discussions apply equally to receiving antennas.

7.2.1 Near-Field and Far-Field Conditions

At distances up to a maximum (about the antenna main dimension) from the antenna aperture, the radiated electromagnetic wave is spherical (i.e., the wave spreads out in a spherical pattern). This is termed the near-field. However, at distances greater than the maximum antenna dimensions, the signal progresses as a plane wave and retains this pattern when it reaches the receiving antenna. This is termed the far-field situation. The antenna will receive signal power and this received signal will progress onto the receive-section circuitry.

Defining the maximum dimension of the antenna as d, then the distance $R_{far\ field}$ following which far-field conditions apply is given by

$$R_{far\ field} = \frac{2d^2}{\lambda} \text{ meters} \tag{7.1}$$

This equation works well for most types of antennas. However, for electrically short antennas including small dipoles and loop designs, (7.1) tends to compute too low and the simplified minimum distance $R_{far\ field} = 2\lambda$ should be applied, where λ is the free-space wavelength.

Example 7.1

Calculate the far-field distance in the case of a 45-cm-diameter satellite broadcast receiving dish reflector antenna. The center frequency of the channel is 14.5 GHz.

First, calculate the wavelength:

$$\lambda = \frac{c}{f} = \frac{3 \times 10^8}{14.5 \times 10^9} = 2.0689 \text{ cm (i.e., 20.689 mm)}$$

Then apply (7.1) to determine the distance at which far-field conditions set in.

$$R_{far\,field} = \frac{2d^2}{\lambda} = \frac{2\times(0.45)^2}{0.020689} = 19.58\,\text{m}$$

In this instance, far-field conditions set in at a distance of 19.58m from the antenna's aperture, which is typically a major dimension of a substantial building down-range.

Example 7.2

Recalculate the far-field distance assuming the same size dish antenna as for Example 7.1 but where the center frequency of the channel is 29 GHz.

Again, first calculate the wavelength:

$$\lambda = \frac{c}{f} = \frac{3\times10^8}{29\times10^9} = 1.03\,\text{cm (i.e., 10.3 mm)}$$

Then apply (7.1) to determine the distance at which far-field conditions set in for this higher frequency example.

$$R_{far\,field} = \frac{2d^2}{\lambda} = \frac{2\times(0.45)^2}{0.0103} = 39.14\,\text{m}$$

A practical maxim is doubling the frequency means doubling the far-field distance.

7.2.2 Radiation Patterns and Beamwidth

As mentioned above, an isotropic antenna (an antenna that radiates with precisely equal field strength in all directions) is purely a theoretical concept and cannot be achieved in practice. The isotropic antenna provides a useful baseline concept against which practical antennas can be compared.

In practice, all antennas are, at least to some extent, directional and a typical antenna will radiate electromagnetic energy in the forms of a main beam plus at least two unwanted forward-directing sidelobes together with several backward-directing sidelobes. This situation is illustrated in Figure 7.2.

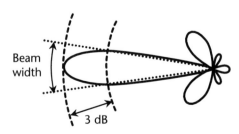

Figure 7.2 Typical antenna radiation pattern.

7.2 Antenna Fundamentals

Figure 7.2 is necessarily a 2-D representation whereas in practice the radiation pattern is 3-D. In this diagram, the beamwidth is also indicated, being the angle subtended by the half-power (or 3 dB) points on the main beam.

7.2.3 Directivity

The directivity of an antenna is a measure of the extent to which the antenna focuses a beam. All antennas possess some specific degree of focusing capability and an equation that expresses the directivity of any pencil-beam-type radiation pattern with fair accuracy is:

$$D \approx \frac{32,400}{\theta_h \theta_v} \tag{7.2}$$

Consider for example a directional antenna for which the horizontal beamwidth is 10° and the vertical beamwidth is 20°. Using (7.2), the directivity is then approximately: $32,400/(10 \times 20) \approx 162$, or in dB $10\log(D) = 22.1$ dB.

Dipoles and other similar antennas are much less directional, tending to have directivities anywhere between 1 and 20 dB.

7.2.4 Radiation Efficiency

Various losses in an antenna (metallic and dielectric) result in an overall internal power loss P_{loss} that reduces the proportion of the total RF input power P_{in} that will actually become radiated. The ratio between the actual radiated power P_{rad} and the signal input power is termed the radiation efficiency η_{rad}, hence:

$$\eta_{rad} = \frac{P_{rad}}{P_{in}}$$

or

$$\eta_{rad} = \frac{P_{in} - P_{loss}}{P_{in}}$$

which is identical to:

$$\eta_{rad} = 1 - \frac{P_{loss}}{P_{in}} \tag{7.3}$$

In (7.3), the term P_{loss} expresses the total power dissipation in the antenna, resulting from resistive losses in the structural materials.

For example, if an antenna suffers dissipative losses amounting to 0.5W and is driven by an input power of 4W, then using (7.3), the radiation efficiency is: $\eta_{rad} =$

$1 - (0.5/4) = 0.875$ (i.e., 87.5%). In general, since losses are always kept as low as possible, the radiation efficiency of most types of antennas is high, typically well over 80%.

7.2.5 Aperture Efficiency

A wide variety of antennas are classed as aperture efficient. This concept infers that aperture efficient antennas exhibit a reasonably well-defined aperture through which the radiation emerges; lensed and array antennas are good examples of this situation.

For these types of antennas the maximum possible directivity D_{max} is given in terms of the aperture area A by:

$$D_{max} = \frac{4\pi A}{\lambda^2} \tag{7.4}$$

In practical situations, however, various features serve to prevent antennas from achieving the maximum possible directivity. Such features include nonideal electromagnetic characteristics, aperture blocking (e.g., by the necessary feed-point electronics), or nonideal feeds and these effects lead to the further concept of aperture efficiency η_{aper}. Thus, aperture efficiency (as a linear factor) must act as a multiplying coefficient on the right side of (7.4) to provide the actual directivity D, resulting in (7.5).

$$D = \eta_{aper} \frac{4\pi A}{\lambda^2} \tag{7.5}$$

in which η_{aper} can never exceed 1 (i.e., this is a per unit efficiency).

Example 7.3

Calculate the diameter of a parabolic reflector antenna operating at a center frequency of 6 GHz, having a directivity of 35 dB and an aperture efficiency of 75%.

First, rearrange (7.5) for aperture area :

$$A = \frac{D\lambda^2}{4\pi\eta_{aper}} \tag{7.6}$$

and with $A = \pi r^2$ where r is the radius of the circular aperture, this becomes:

$$r = \sqrt{\frac{D\lambda^2}{4\pi^2 \eta_{aper}}}$$

or

7.2 Antenna Fundamentals

$$r = \sqrt{\frac{D}{\eta_{aper}}} \frac{\lambda}{2\pi}$$

The free-space wavelength λ at 6 GHz is $3 \times 10^8/6 \times 10^9 = 0.05$m, as a linear quantity the directivity D is $10^{(35/10)} = 3{,}162$ and also as a linear quantity the aperture efficiency η_{aper} is 0.75. Substituting:

$$r = \sqrt{\frac{3162}{0.75} \frac{0.05}{6.283}}$$

giving $r = 0.59$m.

The diameter is twice this (i.e., 1.18m), which is a moderate size of parabolic reflector antenna, suitable for example for a SATCOM ground-based receiver.

7.2.6 Effective Area

In the specific case of a receiving antenna it is vital to establish the proportion of the incoming signal that enters the aperture. The maximum effective aperture area $A_{max,eff}$ can be found from (7.6), but with η_{aper} set to unity, hence yielding

$$A_{max,eff} = \frac{D\lambda^2}{4\pi} \tag{7.7}$$

In which λ is the wavelength at the center frequency of the incoming signal.

7.2.7 Gain

The term gain is often referred to in relation to an antenna, although it can be a confusing misnomer because there is no real power gain involved. Unlike any amplifier that would be specifically designed to increase the power of a signal, with an antenna, there is no DC power supply and no mechanism (e.g., transistors) to increase signal power. However, somewhat unfortunately, the term gain is used in the context of an antenna.

The antenna gain G is really another measure of the directivity and is expressed as

$$G = \eta_{rad} D \tag{7.8}$$

where the radiation efficiency η_{rad} is defined and explained in Section 7.2.4. Equation (7.8) shows that the directivity D is simply multiplied by the radiation efficiency η_{rad} to obtain the gain of the antenna.

In the example used in Section 7.2.3, the directivity is 162, and in the example given in Section 7.2.4 the radiation efficiency is 87.5%. Therefore, combining these two parameters and substituting into (7.8) gives

$$G = 0.875 \times 162 = 141.75$$

or expressed in decibels:

$$G\ (\text{dB}) = 21.5\ \text{dB}$$

which represents an antenna with quite a modest gain.

The great utility of this approach is that power ratios in decibels can be combined with absolute power levels (usually dBm) to evaluate various performance levels throughout practically any RF system.

7.2.8 Equivalent Isotropic Radiated Power

Equivalent isotropic radiated power (EIRP) refers to the main characteristic of an RF transmitter (i.e., the final RF output power and the antenna gain). Since the effectiveness of any RF transmitter depends on the radiated power, EIRP is defined as the product of the output power P_t and the transmitting antenna gain G_t. Equation (7.9) expresses this relationship.

$$EIRP = P_t G_t\ (\text{W}) \tag{7.9}$$

There is another related quantity, known as equivalent or effective radiated power (ERP) and this can be important in practical (especially mobile) communications systems. ERP is closely related to EIRP but referred to the gain of a half-wave dipole antenna, which is 2.2 dB. Therefore:

$$ERP(\text{in dB}) = EIRP(\text{in dB}) + 2.2\ \text{dB} \tag{7.10}$$

Note that this power is expressed in decibels (not dBm) because it is a relative power.

7.2.9 Friis' Equation

In order to plan radio links of any type, it is first necessary to calculate the power P_r received into any antenna's aperture. The conservation of energy requires that the average power density p_{av} an (ideal) isotropic antenna will radiate, at distance l from the aperture, is

$$p_{av} = \frac{P_t}{4\pi l^2}\ (\text{W/m}^2) \tag{7.11}$$

Introducing the fact that practical antennas will have gain (G_t for a transmitting antenna), (7.11) is modified to become:

$$p_{av} = \frac{G_t P_t}{4\pi l^2}\ (\text{W/m}^2) \tag{7.12}$$

7.2 Antenna Fundamentals

When this power density arrives at the receiving antenna, the received power P_r will be given by (7.12) multiplied by the effective aperture area A_{eff} (see Section 7.2.6), giving:

$$P_r = \frac{G_t P_t A_{eff}}{4\pi l^2} \text{ (W)} \tag{7.13}$$

Introducing the gain of the receiving antenna and the fact that $A_{eff} = \frac{\lambda^2}{4\pi}$, the final result (Friis' equation) for the received signal power is obtained:

$$P_r = \frac{G_t G_r \lambda^2}{(4\pi l)^2} P_t \text{ (W)} \tag{7.14}$$

The inverse dependence with the square of link length l is a particularly important feature emerging from (7.14). For example, doubling the link length results in the received power being decreased by a factor of one-quarter. Pozar [2] pointed out this effect is substantially better than the exponential decrease in signal power where a cabled link is implemented instead of a free-space link.

In practical situations, this calculation of received power according to (7.14) is always reduced by various effects such as polarization and impedance mismatches. Impedance mismatch is discussed in the next section.

7.2.10 Impedance Matching

Practical antennas almost never present an ideal matched input impedance (the ideal being most often 50Ω). Instead a matching network has to be interposed between the final (50Ω) feed and the antenna itself. This requirement is shown in Figure 7.3.

The matching network is most often designed and built using passive components (generally microstrip). Chapter 8, in which transistor matching is considered in the context of amplifier design, is also relevant in this respect.

7.2.11 Polarization

The spatial orientation of the electromagnetic fields radiated from an antenna is referred to as the polarization of the antenna. In the case of a simple plane wave

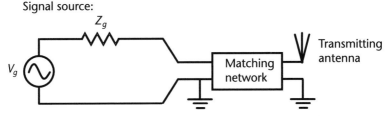

Figure 7.3 A matching network interposed between a signal source and a transmitting antenna ($Z_g \neq 50Ω$).

for which the electric field components exist in only a horizontal or vertical direction, the polarization is linear. When the electric field components are subject to a quadrature phase shift, it results in a counterclockwise rotation as the wave propagates forward in space and time.

This effect is termed circular polarization (CP). Depending on the phasing, this rotation can be left-hand (LHCP) or right-hand (RHCP) as seen directed along the propagation path.

Polarization, notably circular polarization, is important in the design and operation of antennas for many communications applications. It is very important for transmit and receive antennas to have the same polarization and to the same extent. This demands precise physical alignment in order to maintain maximum transferred power. Polarization mismatch refers to any situation in which this alignment is not maintained.

7.2.12 Antenna Noise Temperature

In common with all electronics, so every antenna is a source of noise as well as either radiating or receiving the desired signal. Some of this noise is attributable to the antenna itself (the metal structure in particular) and some actually enters the antenna from the sky, from outer space, from nearby electrical and electronic equipment, and from the ground surface where this is near enough to the antenna. Except for the noise from the antenna itself, all the remaining noise is generally referred to as background noise.

A detailed knowledge of this noise is essential in order to design effective receivers because it enters the receiver subsystem chain (this is discussed further in Chapter 9). Further, antenna noise is directly described in units of temperature (Kelvin). The output noise power N_o is the well-known function of temperature T and system bandwidth B given by (7.15).

$$N_o = kTB \qquad (7.15)$$

where k is Boltzmann's constant = 1.38×10^{-23} J.K^{-1}.

This noise power is equivalent to the thermal noise emerging from a resistor held at temperature T.

It is useful to separate out (equivalently) the background noise and the internal antenna noise sources and this separation is shown in block diagram form in Figure 7.4.

The antenna has physical temperature T_p. The loss due to the attenuator is termed L and is always greater than 1. Also, all the noise brightness and internal (attenuator) noise will be reduced by L because it has to pass through the attenuator. The equivalent noise temperature of this attenuator is $(L-1)T_p$ and the overall antenna noise temperature then becomes:

$$T_a = \frac{T_b}{L} + \frac{(L-1)T_p}{L}$$

or, in terms of the radiation efficiency η_{rad}

7.2 Antenna Fundamentals

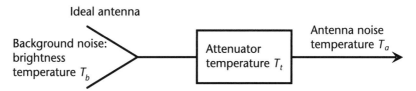

Figure 7.4 Complete antenna noise scenario represented by an ideal antenna receiving brightness temperature T_b followed by an attenuator having noise temperature T.

$$T_a = \eta_{rad} T_b + (1 - \eta_{rad}) T_p \qquad (7.16)$$

Background noise temperature T_b ranges from a few degrees Kelvin through to about 290K depending on the frequency and the antenna elevation angle above horizontal. This noise temperature is lowest (only a few Kelvin) at low frequencies and when the antenna is pointing vertically. It is highest at 60 GHz (about 290K at the O_2 absorption peak) and with the antenna pointing horizontally. Between these extremes the background noise temperature is a highly nonlinear function of frequency.

The overall antenna noise temperatures range from several hundred degrees Kelvin to over 1,000K.

7.2.13 Gain-Temperature Ratio

Gain and noise temperature represent two of the main antenna parameters defined above. Both are of great significance in receiver design and they are often combined to form the ratio G/T_a where G is the antenna gain and T_a is its overall noise temperature. In practice, this ratio is usually simply called the gain-temperature (G/T) ratio and is expressed logarithmically:

$$G/T = 10 \log \frac{G}{T_a} \; (\text{dB/K}) \qquad (7.17)$$

The signal-to-noise input power ratio P_r/N_i is of critical importance for any receiver, and this ratio is directly proportional to G/T. The way in which this fact arises can best be seen by referencing back to Friis' equation for received signal power in (7.14):

$$P_r = \frac{G_t G_r \lambda^2}{(4\pi l)^2} P_t$$

The input noise power N_i is equal to $kT_a B$ so the signal-to-noise input power ratio P_r/N_i can be written as

$$P_r/N_i = \frac{G_t G_r P_t \lambda^2}{kT_a B (4\pi l)^2}$$

Within this equation G_r/T_a can be separated out to rearrange as follows:

$$\frac{P_r}{N_i} = \frac{G_r}{T_a}\left(\frac{G_t P_t \lambda^2}{kB(4\pi l)^2}\right) \tag{7.18}$$

which clearly indicates the direct relationship between input signal-to-noise ratio and G/T. The design importance of this feature is borne out in the next section where dish reflector antennas are considered.

7.3 Dish Reflector Antennas

Reflector antennas implementing parabolic dishes have already been referred to in several of the examples given above (e.g., the penultimate definition in Section 7.1 and Examples 7.1, 7.2, and 7.3).

Microwave antennas based on parabolic dish reflectors are very familiar objects, even to the completely nontechnical observer, because they are encountered in so many locations and applications. The existence of dish-based antennas mounted on rooftops or alongside buildings attest to the fact that millions of homes, globally, are equipped with satellite receivers. Figure 7.5 provides an outline of this type of receiving antenna.

In practice, the receiver comprises a small receiving metallic horn mounted onto a low-noise block downconverter (LNB). Low-noise amplifiers (LNAs) are covered in Chapter 8 and downconverters are described in Chapter 13. Also, in practical installations the receiver is most often mechanically supported by an angled frame secured to the wall of a nearby building (i.e., well away from the concentrating dish).

A parabolic dish reflector causes the electromagnetic wave to become concentrated (or focused) at a point in space in front of the concentrator. The distance of this point away from the face of the dish is of the order of half the diameter.

The receiver, principally the LNB, is located at this focal point so it receives maximum available incoming energy. The output signal from the LNB is at a much lower frequency than the incoming microwave signal, intermediate frequency, which is typically set around 1 to 1.5 GHz as its center value.

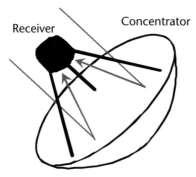

Figure 7.5 Basic concept of a dish reflector concentrator antenna.

Evaluating the gain of parabolic dish reflector antennas is a vital task toward the design of many types of communications systems and a basic formula is derived that enables this calculation. First, it is necessary to combine the equations for directivity (7.5) and for gain (7.8):

$$D = \eta_{aper} \frac{4\pi A}{\lambda^2} \text{ substituted into } G = \eta_{rad} D$$

taking $\eta_{aper} \cdot \eta_{rad} = \eta$ (overall antenna efficiency) and $A = \frac{\pi d^2}{4}$ where d is the physical (circular) diameter of the concentrator, the gain is given by:

$$G = \eta \frac{\pi^2 d^2}{\lambda^2} \tag{7.19}$$

When (as often) the gain is expressed logarithmically, the units are *dBi* (dB relative to the isotropic case).

As described above, in practice, the principal term $\frac{\pi^2 d^2}{\lambda^2}$ will always overestimate the gain of any parabolic concentrator. Practical effects such as overspill of the fields around the perimeter of the dish and some blocking by the feed system (LNB for a receiver) will serve to decrease the gain, but all such effects can be built into the overall efficiency factor η.

Two examples (1-m diameter dish in each case) are: (1) at a center frequency of 14 GHz the gain is 42 dBi; and (2) at a center frequency of 30 GHz the gain is 49 dBi.

From (7.19), it can be seen that the antenna gain increases with the square of the diameter and the inverse square of the wavelength (i.e., the square of the frequency).

Millimeter-wave antennas with diameters around 10m can achieve gains exceeding 70 dBi, which can be required for main transceiver satellite communications ground stations. Such antennas are mechanically complex and require extensive support structures.

Arrays of parabolic dish antennas are also deployed as receivers for deep space communications. Some radar systems also implement parabolic dish antennas, although active, electronically scanned arrays (AESAs) are currently favored (AESAs are described in a later section).

7.4 Flat-Panel or Patch Antennas

For communications systems where the high gain capabilities of dish reflector antennas are not required other (often more basic) types of antenna designs are available. Many of these comprise flat-panel or patch arrays and these are often based on microstrip transmission line configurations (see Section 3.3.2 of Chapter 3). Otherwise, various conducting patch structures are frequently implemented and complete, packaged versions of such antennas take on physical dimensions very similar to those of commercial indoor Wi-Fi hubs.

These types of antennas can provide gains up to about 20 dBi and they are generally installed onto walls indoors. A center frequency of 2.4 or 2.7 GHz usually applies to such antennas and a straightforward flexible coaxial cable provides the output line carrying the intermediate-frequency signal.

Wideband circularly polarized antennas continue to receive a substantial amount of attention relating to patch antennas and related arrays [3]. Circularly polarized (CP) antennas are described in Section 7.2.11. A key aspect of this technology is the axial ratio (AR) of the electromagnetic wave, which is the variation of the magnitude of the electric field versus time. Kovitz et al. [3] investigated this in some detail, notably based on the structure shown in Figure 1.4.

For the array shown in Figure 1.4, the axial ratio (AR) remains below 3 dB over the 1.8-GHz to 3-GHz bandwidth and below 0.5 dB over the 1.8-GHz to 2.7-GHz bandwidth.

7.5 Analog, Digital, and Hybrid Beamforming

Multiple input/multiple output (MIMO) systems are being put forward as potential candidates for several new types of communications systems. Beamforming, both analog and digital, forms a vital part of this approach but because this subject is principally applicable to radars or to prospective 5G systems only a brief overview is provided here. Ghosh [4] provides an excellent overview of the technology.

Analog beamforming is performed at the stage between the (single) transceiver unit and the antenna array as outlined in Figure 7.6.

The function of the analog beamformer, operating at RF, is to dynamically provide different weightings of the RF signal delivered to the elements of the array. The final beam is therefore the combination of all these weightings. This approach provides the best coverage at relatively low cost.

Analog beamforming is often referred to as frequency-flat because no frequency selecting is involved (unlike digital beamforming).

Digital beamforming is performed at baseband (i.e., not RF), as shown in Figure 7.7.

The function of the digital beamformer, operating at baseband, is to dynamically provide different weightings of the signals delivered to the transceivers feeding the array. Again (as with the analog beamformer), the final beam is therefore the combination of all these weightings.

Digital beamforming is often referred to as frequency-selective (unlike analog beamforming). FPGAs (Chapter 5) are used for the digital processing.

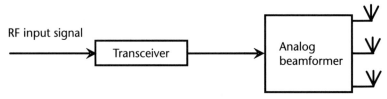

Figure 7.6 Architecture of an analog beamformer.

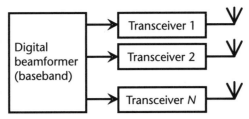

Figure 7.7 Architecture of a digital beamformer.

Hybrid beamforming combines the analog and digital approaches, indicated in Figure 7.8.

With the hybrid beamformer system, adaptive weighting is available in both analog and digital domains. As a result, there is optimization between both coverage and capacity.

Multi-element MIMO arrays are feasible with element counts up to at least 64. These are similar in surface appearance to the 4-element array shown in Figure 1.4 and are typically packaged as described for flat-panel antennas. Operation to at least 30 GHz has been demonstrated.

In manufacturing terms, the 3-D printing of flat-panel antennas also represents an important advancement.

7.6 Active Electronically-Scanned Arrays

Although mainly applying to radars rather than communications systems, the subject of active, electronically scanned arrays is so important that it is discussed briefly here.

An active electronically steered (or scanned) array (AESA) takes the concept of using an array antenna a step further on from the passive electronically scanned phased array (PESA). Instead of shifting the phase of signals from a single high-power transmitter, every AESA employs a grid of hundreds, thousands, or even more transmitter-receiver (TR) modules (or TRMs) that are linked together by high-speed processors.

Each TRM has its own transmitter, receiver, processing power, and a relatively small radiator antenna on top. For some AESAs this radiating element is a microstrip patch, while for others (notably the U.S. AN/APG series) the structure is spike-like. The TRM can be programmed to act as a transmitter, receiver, and/or miniature radar. The TRMs in the AESA system can all work together to create a powerful radar, but they can do different tasks in parallel, with some operating

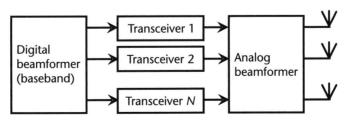

Figure 7.8 Architecture of a hybrid beamformer.

together as a radar warning receiver, others operating together as a jammer, and the rest operating as a radar. TRMs can be reassigned to any role, with output power or receiver sensitivity of any one of the subsystems defined by such temporary associations proportional to the number of modules. The antenna elements are generally spaced at half-wavelength intervals around the array.

An AESA provides 10 to 30 times more net radar capability plus significant advantages in the areas of range resolution, countermeasure resistance, and flexibility. In addition, it supports high reliability combined with low-maintenance operation, which translate into lower life-cycle costs. Since the power supplies, final power amplification, and input receive amplification are distributed, MTBF is substantially higher (typically by a factor of 10 to 100 than that of a passive ESA or mechanical array). This results in higher system readiness and significant savings in terms of life-cycle cost of a weapon system, especially a fighter aircraft.

The use of multiple TRMs also means that failure of up to 10% of the TR modules in an AESA will not cause the loss of the antenna function, but will only marginally degrade its performance. From a reliability and support perspective, this graceful degradation effect is invaluable. A radar that has lost several TRMs can continue to be operated until scheduled downtime is organized to swap the antenna.

Improvement of gallium arsenide (GaAs) material and the development of monolithic microwave integrated circuits (MMICs) have represented key enablers for the development of AESA technology. More recently, gallium nitride (GaN) has been introduced for higher-power RF power amplification with accompanying higher RF conversion efficiency. Several relatively new AESAs implement GaN technology. Driver, phase-shifter, modulator, and amplifying functions are required.

Two prominent early programs in X-band AESA technology development have been the U.S. Army family of radars program (which provided the basis for the X-band AESAs in the THAAD and GBR radars for theater and national missile defense systems, respectively), and the U.S. Air Force programs to produce X-band AESAs for the F-15 and the F-22 fighter jets. The investments in F-35 JSF (Lightning II) radar technology have also fostered pivotal advances by reducing cost, weight, and mechanical complexity. JSF transmit/receive TRMs are referred to as fourth-generation TRM technology.

Practical radars demand upwards of 4W of CW power per element and it is the RF output power requirement that mainly drives up the cost of AESAs. This state-of-the-art AESA technology is becoming the de facto standard for the primary sensor on advanced fighter aircraft and enables even better reliability, reduced life-cycle costs, and improved detection capability.

The amplifying functions fall into three basic categories: driver, power amplifier (PA), and (for the receive section) low-noise (LNA). While GaAs MMICs provide excellent driver and LNA chips, the advent of GaN PAs opens up new opportunities. As GaN PA MMICs gain acceptance and availability, their implementation in AESA TRMs means at least a doubling of radar range.

Figure 7.9 is given here as a very basic outline to aid an understanding as to how an AESA works.

In Figure 7.9 only 30 elements are shown—as small squares on the circular AESA array—whereas in practice this number can be almost anything between 10

7.6 Active Electronically-Scanned Arrays

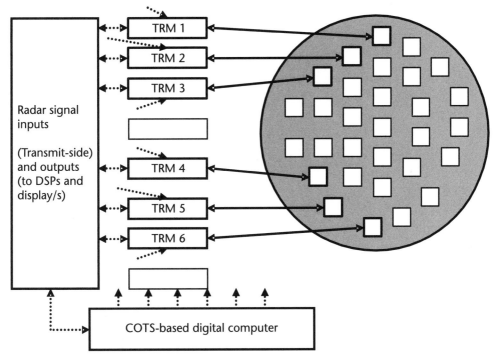

Figure 7.9 Basic, block-schematic outline of an AESA.

and 100,000, or occasionally even more. Also, although shown separately in Figure 7.9, in practice, each antenna element is physically very close to its own dedicated TRM. For simplicity, only six TRMs are indicated in bold outline and labeled. In practice, every one of the 30 elements would need an associated TRM so the total number of TRMs would also be 30. The acronyms COTS stands for commercial off the shelf, DSP stands for digital signal processor, and TRM stands for transmit-receive module. It should be noted that the RF (microwave) signal is generated, modulated, phase-shifted, and amplified on the transmit side of each TRM and processed on both transmit and receive sides.

Each array (antenna) element is inherently associated with a TRM and there are thus N TRMs for a system having N elements. Again, it is important to stress:

$$10 \leq N \leq 100{,}000 \text{ (or very occasionally even more)}$$

The computer controls all the TRMs, turning them on or off by dynamic programming according to the desired real-time scan mode. It is reasonable to assume AESAs will increasingly penetrate both the radars and the communications spaces.

References

[1] Balanis, C., *Antenna Theory, Analysis and Design*, New York: John Wiley & Sons, 2005.
[2] Pozar, D. M., *Microwave and RF Design of Wireless Systems*, New York: John Wiley & Sons, 2001.

[3] Kovitz, J. M., J. H. Choi, and Y. Rahmat-Samii, "Supporting Wide-Band Circular Polarization," *IEEE Microwave Magazine*, July/August 2017, pp. 91–104.

[4] Ghosh, A., "The 5G mmWave Radio Revolution," *Microwave Journal*, September 2016, pp. 22–36.

CHAPTER 8

Small-Signal RF Amplifiers

8.1 Review of Amplifier Fundamentals

The approaches and technologies deployed to design any amplifier vary greatly but many of the components discussed in Chapters 2, 3, and 4 are implemented to create amplifier circuits of various types. CAD and computer-based simulations (EDA) are also extensively used for designing and simulating amplifiers.

The block triangle shown in Figure 8.1 is a basic symbol generally used to represent any amplifier.

This is a type of signal flow representation in which other important aspects such as common earth and DC power supplies are omitted for clarity. The symbol G refers to the power gain of the amplifier. That is, the signal power P_o delivered from the output divided by the signal power delivered into the input P_i as indicated in (8.1):

$$G = P_o / P_i \tag{8.1}$$

When P_o and P_i are expressed in watts or milliwatts, then clearly from (8.1), G is a numeric ratio (i.e., a factor such as 4.3, 27, or 50). However, G is often expressed in decibels rather than as a numeric. Also the input and output power levels are frequently expressed in dBm or occasionally in dBW.

Although amplifier power gain can be defined in various alternative ways (e.g., transduced power gain) for all purposes throughout this book, the relatively simple description given by (8.1) will suffice.

For relatively low-frequency amplifiers, voltage gain is often used. Referring again to Figure 8.1, the voltage gain G_V is defined as signal voltage V_o available at the output divided by the signal voltage V_i available at the input as indicated in (8.2):

$$G_V = V_o / V_i \tag{8.2}$$

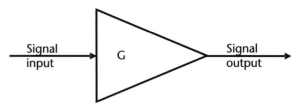

Figure 8.1 Basic block symbol applicable to any amplifier.

However, under RF conditions, the voltages V_o and V_i are virtually impossible to define effectively, which is a major reason why power levels and power gain quantities are usually employed.

The main specification aspects applicable to any amplifier are:

- Obviously achieving the desired power gain over the desired frequency bandwidth;
- Ensuring that the amplifier will be stable (no unwanted oscillations);
- Appropriately applying automatic gain control (AGC);
- Designing for low-noise (with low-noise amplifiers [LNAs]);
- Designing to achieve the desired output power at sufficiently low distortion and high-enough efficiency (with power amplifiers [PAs]).

Other specifications include:

- Gain flatness over the operating band;
- Spectral roll-off characteristics out of band (both ends);
- Input and output VSWR (or return loss);
- Dynamic range;
- Power-added efficiency (PAE);
- Linear phase response;
- Operating temperature limits;
- Duty cycle or peak-to-average power ratio (PAPR);
- Leakage current;
- Multifunctionality;
- Reconfigurability;
- Level of technology maturity;
- Sensitivity to production variations;
- Form factor (package options);
- DC bias supply requirements;
- Lastly, but of paramount importance, cost.

Many of the items included in this list have been taken from those supplied by Kingsley and Guerci in their book [1]. These authors are thanked for their permission for this part of their text to be reused for the above list.

8.2 Basic RF Amplifiers

8.2.1 Practical RF Amplifier Realization

A more comprehensive schematic layout of an RF amplifier is shown in Figure 8.2.

All the components indicated in Figure 8.2 are essentially embodied within the basic symbol of Figure 8.1, with the important exceptions of the DC supply, the associated biasing circuitry, and the lowpass filters. The transistor may be a bipolar device or a field-effect device (in any suitable semiconductor material).

The input matching network may be designed for noise-matching (Chapter 9) or for maximum RF power transfer in the case of a power amplifier (Chapter 10). The output matching network is almost always designed for maximum RF power transfer to the load (i.e., to the next stage).

The DC supply voltage rail is typically 3V for most portable systems or as high as 20V or even 40V for GaN HEMT-based power amplifiers. The DC biasing circuitry is designed so that the appropriate DC bias is fed to the transistor input terminal and also (a different value) to the transistor output terminal. In each instance, lowpass filters ensure that most RF energy is not transferred back into the DC supply.

Only a relatively small amount of RF energy returns to the DC supply, and this supply is effectively decoupled from the RF path (hence, only a low level of loading).

The simplest practical realization of the lowpass filters comprises inductances (L), which effectively block the RF signal. Such components are sometimes termed DC blocks or RF chokes (RFC). However, in practice, considerably more attention must be given to the design of the DC biasing circuitry and an array of capacitors must be included to further shunt RF signal components to ground, often over a wide range of frequencies (broadband). A typical two-stage circuit is shown in Figure 8.3.

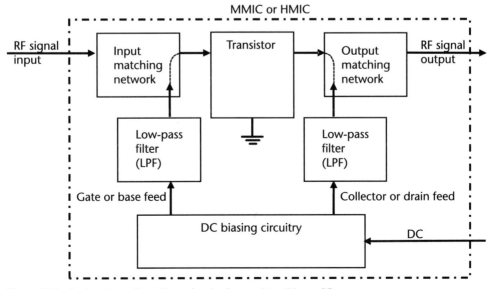

Figure 8.2 Basic schematic outline of a single-transistor RF amplifier.

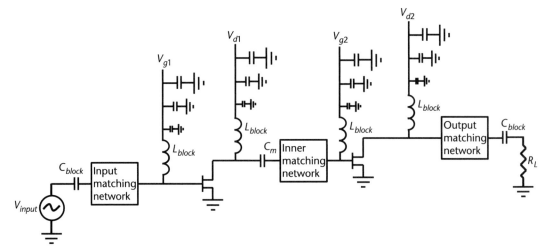

Figure 8.3 Two-stage amplifier with individual gate and drain bias. (© Artech House 2016.)

As shown in Figure 8.3, in practice, a series of decoupling capacitors must be included as well as the series chokes L_{block} between the DC supply and the transistor (or MMIC). The reason for this is to prevent any unwanted signals that may exist of the (supposedly pure) DC supply from entering the RF sections of the circuit. It is usually advisable to ensure that these capacitors cover a very wide range of frequencies and, as a result, at least three capacitors may be required, the first accommodating lower frequencies (the largest value, several microfarads), a second handling moderate frequencies (a fraction of a microfarad) and a third capacitor offering typically several picofarads.

8.2.2 Interstage or Inner Matching Networks

A major initial requirement for any amplifier (not just RF amplifiers) is to achieve sufficient power gain (G) to meet the system specification. Often, this requirement cannot be met using one transistor alone and at least two transistor stages are needed. When two transistors are to be connected in series (more precisely in cascade) to meet this requirement, then interstage networks (or inner matching networks; see Figure 8.3) must be implemented as shown in Figure 8.4.

Like Figure 8.1, this again is a type of signal flow representation in which other important aspects such the DC power supply rail are omitted for clarity. This interstage requirement is especially important in gain blocks and driver amplifiers.

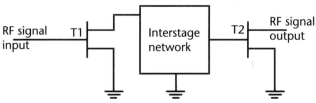

Figure 8.4 Block schematic indicating an interstage network interposed between two amplifying transistors in cascade.

The design of interstage networks (or inner matching networks) is a necessary procedure and such networks comprise passive circuitry interposed between transistors with (usually) the aim of maximizing power transfer. The transistors must have sufficiently high transition frequencies (f_T) to meet the upper frequency bound with the bandwidth. f_T is defined in Chapter 2.

Throughout most of this chapter, it is assumed that the amplifiers are all operating under linear small-signal conditions. The term small-signal means that all signal levels are very much smaller than the bias voltages and currents used to power and set operating conditions.

In practice, the technology of choice for amplifiers (other circuits also) is either integrated or hybrid circuits based on discrete transistors.

DC supply rail voltages vary greatly, being nominally 3V for portables including mobile phones. However, the supply voltage required for a power amplifier implementing GaN HEMTs usually lies somewhere in the range from 20V to 40V.

8.3 The Vital Issue of Stability

In this context by stability is meant a design that is intended to be an amplifier and only an amplifier may for some reason actually undesirably oscillate at a particular frequency (or frequencies). In theory (often in practice), any amplifier design may be prone to such instability and the fundamental concept indicating how this instability arises is exhibited in Figure 8.5.

In Figure 8.5, the amplifier is shown delivering the output signal to the load that has impedance Z_L. The important point to observe is the inherent existence of an internal mainly capacitive element effectively connecting between output and input signal ports. (Among the small-signal scattering [S] parameters this tendency for instability is inextricably linked to the S_{12} reverse transfer parameter.) The term stability here refers to design for the avoidance of unwanted oscillations.

The first point to note is that any amplifier, improperly designed, can easily turn into an unwanted oscillator (i.e., generating its own undesired signal).

This is an increasingly serious problem as both open-loop gain and operating frequencies increase. Therefore, at microwave or millimeter-wave frequencies, the issue of stability must be addressed.

The fundamental reason for the potential instability problem is that a perfectly unilateral amplifier cannot exist in practice. (In electronics, a unilateral circuit is one in which there is absolutely no electrical connection between the input and output internal terminals.) Instead, all amplifiers possess internal feedback (Figure

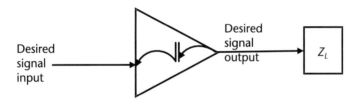

Figure 8.5 Schematic circuit diagram indicating how instability can arise.

8.5) and therefore a portion of the output signal becomes fed into the input (along with the wanted signal).

In practice, it is vital to check whether any proposed design may be subject to potential instability and the complete Nyquist stability test is built into modern simulators. An example of a test for stability is the one provided by Campbell and Brown [2]. In their paper on the S-probe approach, Campbell and Brown indicated two very important equations, which, when proven to coexist simultaneously, will definitely show the amplifier is unconditionally stable. These equations are for the stability indices S_1 and S_2:

$$S_1 = Re(\Gamma_{in}\Gamma_s) < 1 \tag{8.3}$$

and

$$S_2 = Re(\Gamma_{out}\Gamma_L) < 1 \tag{8.4}$$

Note that S_1 and S_2 are not S-parameters.
In (8.3) and 8.4):

- Γ_{in} is the reflection coefficient applying to the input port of the amplifier.
- Γ_s is the reflection coefficient applying to the signal source for the amplifier.
- Γ_{out} is the reflection coefficient applying to the output port of the amplifier.
- Γ_L is the reflection coefficient applying to the load attached to the output of the amplifier.

All of these reflection coefficients are exactly related to the S-parameters of the amplifier (or of each transistor stage) and to the source and load impedances.

Provided that the conditions of (8.3) and (8.4) hold, there is no possibility of oscillation at any frequency, that is, the circuit is absolutely and unconditionally stable. This will remain the case under all worst-case source and load conditions.

However, two factors should be borne in mind:

- In each case, for S_1 and S_2 in (8.3) and (8.4), the requirement is for the results to be less than 1.
- The S-parameters as well as the source and load impedances must be known very accurately.

In the case of the first requirement, there could be a serious practical issue if in a particular case the calculation came out very marginally less than 1 (e.g., 0.98). The problem with this situation is that in the event of a slight change in any parameter the inequality could easily exceed unity and the circuit would no longer be absolutely and unconditionally stable. This could happen, for example, as a result of a batch variation, a temperature change, or a bias alteration.

Most EDA software suites include a subroutine that embodies a rigorous stability check and NI AWR, for example, offers what the company term the GPROBE2 element in order to check for stability. GPROBE2 is fundamentally based on the above approach to analyzing for stability.

It is extremely important to guard against this unwanted oscillation. As far as RFICs or MMICs are concerned (Chapter 2), it is essential to ensure that designs offer (ideally) unconditional or at least conditional stability. If the specifications of such chips are being examined with a view towards possible purchase, then it is highly recommended that only products specifically stated as being unconditionally stable are chosen. (Conditional stability means the circuit will remain stable, but when and only when limitations usually specified by the manufacturer are observed.)

8.4 Fundamental Receiver Characteristics Leading to the Need for AGC

Where substantial receiver gain is required of a gain block this is normally (and best) accommodated at intermediate frequency (IF) rather than RF. While the IF may be around 100 MHz in many wireless receivers (e.g., cellular, mobiles), it can be many gigahertz in higher-frequency microwave or millimeter-wave systems.

Implementing as much as possible gain at IF has at least two major advantages:

- It is easier to stabilize (see Section 8.3 regarding amplifier stability).
- It tends to be cheaper than at RF/microwave (it can be much cheaper).

Many (indeed probably most) radio and radar receivers exhibit the following characteristic: ~100 dB or more of total through power gain.

However, a potentially serious problem is encountered when dynamic ranges are considered. This is typically around 80 to 100 dB at the receiver input, but typically around 60 dB at the receiver output.

Now assume that the receiver has the fairly modest fixed total power gain of 80 dB. It is worthwhile calculating this as a linear gain factor (it is a factor of 100 million); and then consider two possible extremes of input power levels and the attempted resulting output power levels:

The attempted output power is:

1. A relatively low-level input signal of 10 nW: 1W (probably acceptable);
2. A relatively high-level input signal of 10 μW: 1 kW.

In the second case, the receiver would probably explode or at least severe limiting would occur taking the amplifier into a highly nonlinear output mode.

Because of this clear problem, it is essential to have some form of AGC. The receiver must have more gain for low-level input signals (perhaps 70 or 80 dB) and then much less gain as the input signal power increases. This is performed at the IF for the reasons given above.

8.4.1 Toward an Effective AGC Circuit Design

From the above discussions, it can be seen that a special circuit is needed in order to automatically control the overall gain through the receiver chain. Toward this aim, a classic approach is to consider some form of active feedback and an overall outline of this approach is indicated in Figure 8.6.

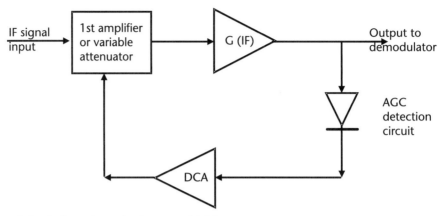

Figure 8.6 Outline schematic of proposed AGC arrangement.

In Figure 8.6 (also all relevant circuits, discussions and equations throughout this book), IF means intermediate frequency. For signal processing convenience in many receivers, the entire band of signal frequencies is often downconverted to an IF. The center frequency of an IF band is usually set at some standard value, such as 470 MHz in many microwave systems. In Figure 8.6, the main IF amplifier G (IF) will have the highest power gain of any component within the system; it could be as high as 40 dB or even greater. The design concept behind the outline schematic of Figure 8.6 is to tap off a relatively small sample of the signal, convert this sample to DC (produced by the diode as shown), apply this relatively slowly changing DC to a voltage amplifier (DCA), and then apply the amplified voltage to the first stage through which the original IF signal passes: the first amplifier or variable attenuator. Therefore, when the output power level from the main IF amplifier increases, this is sampled, detected by the diode, and applied to the voltage amplifier so that either the gain of the first amplifier is automatically decreased or the attenuation of the variable attenuator is increased. Either way, the signal power is dynamically reduced as required.

A practical realization of this type of arrangement is shown in Figure 8.7.

Schottky diodes are described in Chapter 2 and lowpass filter (LPF) technologies are described in Chapter 6. Following the Schottky diode detector, only the DC level is required to be applied to the inverting input of the operational amplifier A. The capacitor C (usually several picofarads) shunts any signal current components to ground. DAC is the standard abbreviation for digital-to-analog converter (Chapter 11).

- The diode-capacitor combination (D, C) is the AGC detecting circuit.
- The voltage amplifier (A) must have moderate voltage gain but low bandwidth; it must be open-loop stable. Its purpose is to amplify the difference voltage indicated by:

$$\left|V_{AGC} - V_{Ref}\right| \tag{8.5}$$

8.5 High-Gain RF Amplifiers

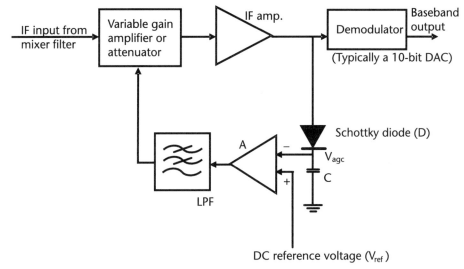

Figure 8.7 Details of a typical AGC circuit.

- A variable attenuator is often used; there will be very little noise effect due to this because it is connected far down the receiver chain. The effects of noise are considered in Chapter 9.

For a hybrid module realization of this circuit all the components are either readily available off-the-shelf or easily created on the circuit board. Practically any commercially available operational amplifier could be selected for the voltage amplifier A. Alternatively, all the sections could be realized either as a stand-alone MMIC/RFIC or even as part of a more complex RFIC.

Frequency-band conversion (mixing, the meaning of IF and related information) is dealt with in Chapter 13.

8.5 High-Gain RF Amplifiers

As an example of high-gain narrowband amplifier design, consider a single-FET-based configuration that is required to operate at a center frequency of 2.4 GHz. Design aspects followed through in this section proceed manually (i.e., step-by-step expressions and calculations rather than resorting to CAD approaches). This is so that the reader can gain detailed insights into important design aspects, including the implementation of microstrip transmission lines.

The transistor has the following measured parameters (all at 2.4 GHz):
Input admittance $G_{in} - jB_{in}$:

$$Y_{in} = 0.12 - j0.08 \tag{8.6}$$

and

$$G_{a,max} = 13 \text{ dB} \tag{8.7}$$

where $G_{a,\max}$ is the maximum available power gain. To achieve maximum output power, this transistor must be impedance matched.

The first requirement is to understand that the (narrowband) input matching network is often formed using an L-section of reactances. There are two alternatives: (1) a series inductance followed by a shunt capacitance, or the reverse, that is, (2) a series capacitance followed by a shunt inductance. The first alternative is often chosen because DC bias can be fed through the series inductance.

In practice, the reactances resulting from the inductors and capacitors will often be realized as two interconnected microstrip transmission lines. The challenge is to determine the widths and lengths of each of these lines operating at the 2.4-GHz center frequency. This operation must be undertaken for both the input and the output matching and the general arrangement is shown in Figure 8.8.

In Figure 8.8 the source, output load, FET and their connections are shown symbolically.

There would often be a short length of 50-Ω microstrip line connecting between the source and the initial series matching element but this is not shown in Figure 8.8. This line would be labeled A so that the remaining sections of microstrip matching elements are termed B and C, respectively.

Since the series matching element is a quarter-wavelength impedance transforming section of the microstrip line, matching the transistor input conductance (0.12 S). From the basic transmission line theory, the characteristic impedance of this line is given by:

$$Z_{0B}^2 = Z_0 / G_{in} \tag{8.8}$$

refer to (8.6) in relation to this, that is, in this case:

$$Z_{0B}^2 = 50 / 0.12 \tag{8.9}$$

Therefore:

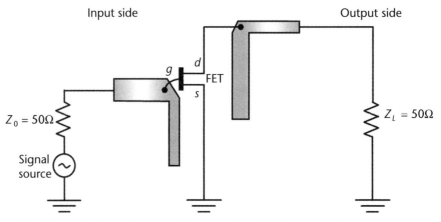

Figure 8.8 Schematic circuit diagram of the narrowband FET-based amplifier.

$$Z_{0B} = 20.4 \Omega$$

The shunt section of microstrip line must tune out the input (capacitive) susceptance of the transistor ($-j0.08$ S); see (8.6). Therefore, this section of line must be set inductive and again applying basic transmission line theory this is accommodated with an eighth-wavelength line shorted to ground at its extremity. The input impedance to this section of line (C) is given by:

$$Z_{inC} = jZ_{0C}\tan(2\pi l/\lambda_g) \quad (8.10)$$

where Z_{0C} is the required characteristic impedance and λ_g is the wavelength in this section of line, length ℓ. Substituting and canceling quantities leads to:

$$Z_{inC} = jZ_{0C} \quad (8.11)$$

which must be equated with the input susceptance $-j0.08$ S, that is,

$$-j0.08 = -j\frac{1}{Z_{0C}} \quad (8.12)$$

which immediately leads to the result:

$$Z_{0C} = 12.5 \Omega$$

Now that the characteristic impedances of each line are known (20.4Ω and 12.5Ω), the physical dimensions of each section of line can be calculated.

Microstrip design is described in some detail in Chapter 3 and for EDA the relevant equations are incorporated within the algorithms. Alternatively, the online calculator provided by Microwaves101 could be used in this regard [3].

It is assumed the circuit will be constructed on a polymer-based circuit card (substrate) have a relative permittivity of 2.3 and an average thickness of 0.5 mm. Dispersion will be negligible at the frequency of 2.4 GHz. These quantities form the input data.

The output data from this calculator include effective microstrip permittivity values, microstrip widths, and wavelengths. The widths form one aspect of the required results and the line lengths are determined using fractions of the wavelengths. Therefore, for lines B and C, $w_B = 4.85$ mm, $l_B = 22.1$ mm, $w_C = 6$ mm, and $l_C = 10.4$ mm.

Alternatively line C could be made three-eighths of a wavelength long rather the one-eighth implemented in the above design. This would make $\ell_C = 31.2$ mm, although this value may be too large for the available area.

A totally different technology option (replacing line C) could be to go for a lumped discrete inductor of susceptance $+j0.08$ S. At the center frequency of 2.4 GHz, this would mean an inductance of 0.83 nH and such an inductor can be realized in the form of any one of the structures described in Chapter 3.

The output matching network would be designed using similar principles to those described above for the input network. However, the output admittance of the FET and the load impedance would first have to be known.

8.6 Broadband Amplifiers

8.6.1 Basic Requirements

So far in this chapter, the focus has been on relatively narrowband amplifier design, typically up to a few percentage of bandwidth. However, broader bandwidth systems are increasingly important, some even DC to 50 GHz or more, and suitable amplifier designs are vital.

In many instances, broadband amplifiers can be designed using multielement networks in which the entire structure of an LC ladder network is designed to maximize the match to the active device. The series inductances (L) ensure that DC can be readily supplied to the gate of the first transistor. It is essential these matching networks possess relatively low Q-factors so the impedance transformation remains close to the real line of the Smith chart (i.e., corresponding to R + j0 Ω). Often the impedance to be matched differs by as much as an order of magnitude from the source impedance (usually 50Ω). In the following example the impedance to be matched is 5Ω. The center frequency is 10 GHz and the procedure is shown in Figure 8.9 for a two-element network and also for a seven-element network. It is best to run the procedure with the aid of a Smith chart, which is easily called out under practically any EDA (CAD) design algorithm (e.g., ADS, NI AWR).

The two-element network exhibited a 10-dB return loss bandwidth of 2.4 GHz (i.e., a 24% bandwidth centered on 10 GHz), which is a substantial bandwidth that would be adequate for some applications. However, the seven-element matching circuit delivered a 10-dB return loss bandwidth of 11.7 GHz (i.e., a 117% bandwidth centered on 10 GHz). This is a very wide bandwidth, which would well exceed that required for many applications.

There are several contrasting alternatives for amplifier design strategies, and two of these are now presented.

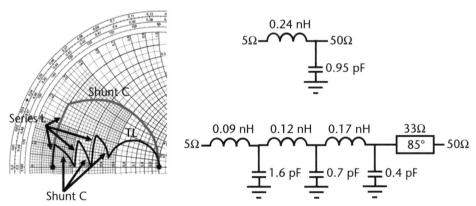

Figure 8.9 Two-element and seven-element matching circuit transformations (5Ω to 50Ω) [1]. (© Artech House, 2016.)

8.6.2 Balanced Amplifiers

The basic idea behind this radically different approach to RF amplifier is that, by implementing two nominally identical internal amplifiers working in conjunction with hybrid (or Lange) couplers, there will be an effective cancelation of the unwanted reflected signals. The overall concept is illustrated in Figure 8.10.

For convenience, assuming voltages rather than signal power levels, then the voltage directions diagrams shown in Figure 8.11 indicate how the reflected signal cancelation works in a balanced amplifier.

Initially the input signal is shared equally (V/2) between the two internal amplifiers, although with a 90° phase difference caused by the couplers. Signals are reflected from the amplifier inputs determined by their input reflection coefficients Γ_1 and Γ_2. Then, provided that these reflection coefficients are equal ($\Gamma_1 = \Gamma_2$), the reflected voltages appearing at the input port will differ in phase by 180° and they will cancel. The greater part of the reflected power goes into the resistive termination on the isolation port (lower LHS of the diagram in Figure 8.11). The technology of this resistive termination must be selected so as to cope with the power

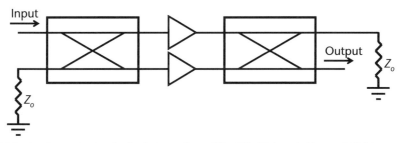

Figure 8.10 A schematic circuit of a balanced amplifier [1]. (© Artech House, 2016.)

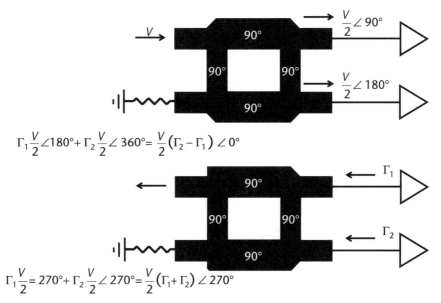

Figure 8.11 System of reflected voltages around the input side of a balanced amplifier [1]. (© Artech House, 2016.)

entering into it at this port. In practice the return loss of each internal amplifier should not exceed 5 dB [1].

Although the balanced amplifier requires two internal (nominally identical) amplifiers, it provides substantial advantages for system designers. In practice, Lange couplers tend to be used (see Chapter 4). High-power balanced amplifier designs increasingly implement GaN HEMT-based internal amplifiers (more details are provided in Chapter 10).

For medium-power, relatively low-power, or low-noise balanced amplifiers, MMIC realization is a common method. The Lange couplers designed within a MMIC balanced amplifier have their respective finger elements interconnected by means of air-bridge conductors.

8.6.3 Distributed Amplifiers

A distributed amplifier is a type of traveling-wave amplifier in which bandwidths upward of decades can be achieved. The concept is based around the interconnection of a cascade of transistors connected between lengths of transmission lines. However, this concept dates to 1948 [4] when vacuum tubes were the only amplifying devices available. The transistor was invented in 1947 and the wideband microwave amplifying device known as the traveling-wave tube (TWT) was invented by Kompfner in 1948. In the TWT, the signal to be amplified is fed into what is termed a slow-wave structure (a metal helix or multiridged waveguide) so that the linear component of its velocity closely matches that of the electron beam, which passes within this slow-wave structure. In this way, amplification is achieved as the signal progressively extracts energy from the electron beam. Over half a century later, microwave (or millimeter-wave) transistors replace the electron tube for the realization of distributed or traveling-wave amplifiers.

In the following design outline, parts of the paper by Bowers and Riehl [5] have used as a basis, starting with a basic schematic diagram of a three-stage distributed amplifier (Figure 8.12). A greatly simplified FET equivalent circuit is employed in Figure 8.13.

In order to match the phase shifts on both the gate and drain lines as well as matching the impedances to 50Ω, it is necessary to impose special conditions on the circuit capacitances; [5] should be consulted regarding this aspect.

The impedance of the gate and drain lines are well-matched to 50Ω for frequencies well below the cutoff frequency of the transmission line. This design approach can be improved by optimizing the lines for phase shift per segment and impedance. In practice, to achieve substantial gain, more stages (sections) are necessary, usually around or exceeding 7. In this design example, 10 stages are implemented and the lumped-equivalent component values (Figure 8.12 nomenclature) are: L_g = 0.4 nH, C_{ga} = 0.32 pF, L_d = 0.41 nH, and C_{da} = 0.06 pF.

This nascent amplifier must be computer-optimized in order to take the design to the next level. Further, the optimization must account for the drain source resistance (R_{ds}, in parallel with C_{ds}) and the gate source input resistance (R_i, in series with C_{gs}). Bowers and Riehl [5] provide details of the optimization process in which a minimization function is introduced (for the inductances and capacitances listed above), equating a summation of the gate and drain reflection coefficients as well as the real parts of the gate and drain electrical lengths.

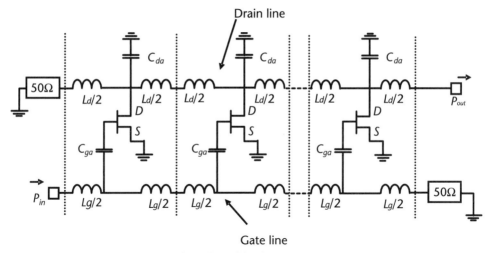

Figure 8.12 Basic three-stage distributed amplifier layout.

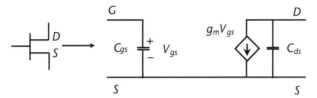

Figure 8.13 Greatly simplified FET equivalent circuit model [7].

The optimal number of sections to use can be computed at the target mid-band frequency from an expression given by Pozar [6]

$$N_{opt} = \frac{\ln(\alpha_g l_g / \alpha_d l_d)}{\alpha_g l_g - \alpha_d l_d} \tag{8.13}$$

where the loss per section on the gate and drain lines, respectively, is given by (8.14) and (8.15):

$$\alpha_g l_g = -Im\theta_g \tag{8.14}$$

$$\alpha_d l_d = -Im\theta_d \tag{8.15}$$

Following this design approach, distributed amplifiers with tens of gigahertz bandwidth and at least 20-dB power gain can be realized.

Distributed amplifiers may be designed in hybrid MIC format or they may be realized in MMIC format, and one example is provided in Figure 8.14.

In this image, the wiggly microstrip transmission lines can clearly be seen. The gate-side connections are through the thinner (higher impedance) line while the drain-side connections are through the thicker (lower impedance) line. It is

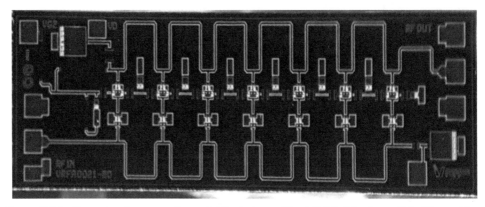

Figure 8.14 Example of a seven-stage distributed amplifier MMIC chip (bare die). (Simon Chan of Viper RF provided this image. Viper RF are thanked for their permission to reproduce the image.)

important to appreciate that each line will be subject to dispersion which is a complication on design. This effect will be particularly significant at frequencies above around 10 GHz and this amplifier is designed to operate at frequencies up to 45 GHz (the impact of dispersion is discussed in Chapter 3).

These types of amplifiers exhibit performance specifications including:

- Over 20 dBm of output power over the full bandwidth;
- An RF power gain exceeding 15 dB over the full bandwidth;
- Peak-to-peak output voltages of several volts into a 50-Ω load;
- Input and output ports both fully matched to 50Ω.

The DC supply voltage (V_{dd}) is typically 7V and the current drawn (I_{ds}) is several hundred milliamperes.

It is interesting to observe that this amplifier would easily cover the initially proposed 5G band centered on 28 GHz (nominally covering 27 to 29.5 GHz).

With any proposed amplifier design, it is always vital to check as to whether there is any possibility of potential instability and (as mentioned in Section 8.3) the complete Nyquist stability test is built into modern simulators. An important aspect to appreciate is that distributed amplifiers are unsuitable for the realization of efficient power amplifiers. RF power amplifiers are the subject of Chapter 10.

References

[1] Kingsley, N., and J. R. Guerci, *Radar RF Circuit Design*, Norwood, MA: Artech House, 2016.

[2] Campbell, C., and S. Brown, "Modified S-Probe Circuit Element for Stability Analysis," (exact reference and date unclear but most likely IEEE, MTT and around 1994 or 1995. S_Probe_White_Paper.pdf, http://kb.awr.com/display/HELP/Modifed+Gamma+Probe+%28GPROBEM%29+Reference+Paper#gsc.tab=0.

[3] https://www.microwaves101.com/calculators/1201-microstrip-calculator.

[4] Ginzton, E. L., et al., "Distributed Amplification." *Proc. IRE*. Vol. 36, August 1948, pp. 956–969.

[5] Bowers, K., and P. Riehl, "Broadband Microwave Distributed Amplifier," University of California at Berkeley, Department of Electrical Engineering and Computer Science, Class Projects, May 21, 1999, http://www-inst.eecs.berkeley.edu/~ee217/sp03/projectsSP99/Dist_Amp_b+r.pdf.

[6] Pozar, D. M., *Microwave Engineering*, 2nd ed., New York: John Wiley & Sons, 1998.

CHAPTER 9
Noise and LNAs

9.1 Introduction

The first question that must be answered is: What exactly is noise, in the electrical context? Perhaps the immediate answer is something like: unwanted and irritating sounds that tend to ruin or even swamp received audio signals, whether musical or vocal.

However, the concept of noise extends way beyond only the audio example. In the broadest sense, noise comprises unwanted and apparently random perturbations that can also damage video content or cause errors in digital signals. Electrically, noise power covers a very wide spectrum or alternatively can be interpreted as an extremely messy perturbation in terms of voltage in the time domain. Both aspects are illustrated in Figure 9.1.

It is not possible to call Figure 9.1(b) a waveform because being random it certainly does not possess any kind of form. A good, brief summary of noise and its effects in electronic systems has been provided by Browne [1]. Further information is available from the books by Pozar [2] and Kingsley and Guerci [3].

Noise cannot ever be completely eliminated, but its effects can and must be minimized. The aim of this chapter is to:

- Define and analyze some important aspects of noise;
- Identify the sources and effects of noise in transistors;
- Provide some state-of-the-art data as guidance;
- Develop the basis for designing LNAs.

Fundamentally, noise power (N) depends on temperature (T, in Kelvin), bandwidth (B), and Boltzmann's constant ($k = 1.38 \times 10^{-23}$ J.K^{-1}):

$$N = kTB \tag{9.1}$$

and this result has the fundamental dimensions of power in watts.

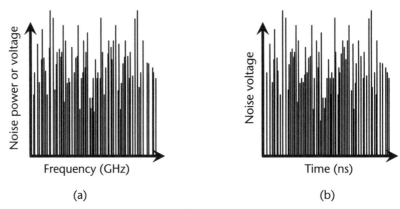

Figure 9.1 (a) Random noise spectrum and (b) time-domain expression.

However, there are several distinct sources of noise, which will be discussed later here.

9.2 Noise Factor, Noise Figure, and Equivalent Noise Temperature

At any point in a communications (or radar) system, the signal-to-noise power ratio (S/N) is highly significant. Furthermore, the ratio of ratios comparing input S/N to output S/N is also of great importance and is called the noise factor (F). With subscripts i meaning input and o meaning output, this ratio of ratios can be expressed as:

$$F = \frac{S_i/N_i}{S_o/N_o} \tag{9.2}$$

Note that this important result is often incorrectly referred to as the noise figure, whereas the noise figure is strictly expressed in decibels $(F_{dB}) = 10\log F$ (in which F is defined by [9.2]).

In the case of an amplifier, these quantities can be identified from Figure 9.2 (this is an annotated version of Figure 8.1 in Chapter 8).

Notice that not only is the (required) signal amplified but the amplifier also contributes further noise of its own. This represents a key feature that particularly affects receiver systems, and therefore LNA design is mainly directed toward minimizing the noise contributed from within its own electronics.

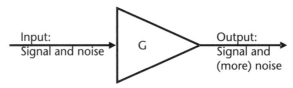

Figure 9.2 Amplifier with input and output signals and noise.

9.2 Noise Factor, Noise Figure, and Equivalent Noise Temperature

Returning to (9.2) and remembering that S_o/S_i is the power gain of the amplifier (G), (9.2) can also be rewritten as:

$$F = \frac{N_o}{GN_i} \quad (9.3)$$

Assuming that all the input noise power kTB is amplified and further that the noise contributed by the amplifier can be expressed in terms of a quantity termed the equivalent noise temperature (T_e) (which is referred to the input), then (9.3) expands to form:

$$F = \frac{GkTB + GkT_eB}{GkTB}$$

The quantities G, k and B all clear through numerator and denominator of the right side of the expression, leaving the result:

$$F = 1 + \frac{T_e}{T}$$

which is easily rearranged (with T substituted as T_s, the standard or reference temperature); the final result for equivalent noise temperature T_e is:

$$T_e = (F-1)T_s \quad (9.4)$$

(T_s is usually taken as the IEEE standard temperature of 290K.)

Remembering that in this expression F is the noise factor (linear) and not the noise figure substituted in decibels. A realistic numerical example is given next.

Example 9.1

An LNA has a power gain of 18 dB and a mid-band noise figure of 2 dB. If the standard temperature is taken as 300K, what is the equivalent noise temperature?

First, from the noise figure the noise factor (linear) must be calculated using $F_{dB} = 10 \log F$, which gives:

$$F = 10^{F_{dB}/10}$$

Substituting the given values yields $F = 1.585$ and then using (9.4):

$$T_e = (1.585 - 1).300$$

or

$$T_e = 175.5K$$

While noise figures (dB) are extensively used to characterize most LNAs and more comprehensive receiver subsystems, equivalent noise temperatures are employed for particularly sensitive receivers (such as many spacecraft systems) or for radars in several instances. However, very importantly, the concept of equivalent noise temperature is particularly useful for understanding and analyzing the noise figure of an attenuating element and the minimum detectable signal for a receiver subsystem.

9.3 Noise Figure for an Attenuating Element

This represents a special case regarding the effect of noise generation. The result is often regarded as being somewhat counterintuitive and the analysis is provided here. Consider an attenuator (or other lossy element) as indicated in Figure 9.3 and with a resistor value R at temperature T.

The attenuator will include its own noise power (N_{extra}) so that the total output noise power is:

$$kTB = \frac{1}{L}(kTB + N_{extra})$$

Solve for N_{extra}:

$$LkTB = kTB + N_{extra}$$

or

$$N_{extra} = (L-1)kTB \qquad (9.5)$$

But also:

$$N_{extra} = kT_e B$$

where T_e has been defined above. Equating this to (9.5) yields:

$$T_e = (L-1)T$$

Applying (9.4) and solving for noise factor F yields:

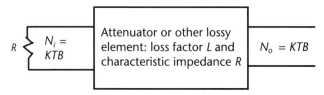

Figure 9.3 Noise analysis of attenuator (or other lossy element).

$$F = 1 + (L-1)T/T_s \qquad (9.6)$$

Example 9.2

Calculate the noise figure (dB) for an element that provides 6 dB of attenuation. Assume that the element is at standard temperature, 300K.

First, the linear factor (L) must be calculated from the 6-dB value. The result is 3.981. Substituting this quantity into (9.6) and with $T = T_s$ gives the noise factor:

$$F = 1 + (3.981 - 1) \text{ or } F = 3.981$$

The noise figure is then $F_{dB} = 10\log F$. In this example, $F_{dB} = 10\log 3.981 = 6$ dB.

This somewhat circular calculation proves that the noise figure of a passive network is equal to its loss (always). In practice, this is a very high noise figure, which would substantially challenge the overall receiver designer. It will be seen later here that any attenuator like the one in this example should be placed several blocks down the receiver chain.

9.4 Minimum Detectable Signal

Every reliable communications receiver requires the incoming RF signal to be at or above a minimum level in order that the system will operate properly. This signal (power) level is termed the minimum detectable signal (MDS). A concept of MDS is provided in the spectrum shown in Figure 1.6.

Much of the following analysis broadly follows the work of Pozar [2]. The effects of MDS can be expressed in terms of the following ratio:

$$\frac{signal + noise + distortion}{noise + distortion} = \frac{S + N}{N} = 1 + \frac{S}{N} \qquad (9.7)$$

In which the (noise + distortion) combination is lumped together as noise power (N). This ratio is sometimes called the SINAD of the receiving system.

Now, the receiving antenna has a characteristic noise temperature T_a (see Chapter 7) and this antenna noise will be fed to the input of the receiving electronics, together with the input signal noise S_i. Combining earlier equations for the noise factor (involving the signal gain G), the total output noise power is

$$N_o = kBG(T_a + T_e) \qquad (9.8)$$

in which all terms have their previously defined meanings. In a superheterodyne receiver, the bandwidth B would be determined by the IF bandpass filter. In many modern systems, B will be controlled by an initial (often tunable) bandpass filter. Dealing with minimum signal power levels and upon introducing the output noise power N_o, the MDS will be

$$S_{i_{min}} = \frac{S_{o_{min}}}{G} = \left(\frac{N_o}{G}\right)\left(\frac{S_o}{N_o}\right)_{min} = kB(T_a + T_e)\left(\frac{S_o}{N_o}\right)_{min}$$

Using (9.4), this relationship becomes

$$S_{i_{min}} = kB[T_a + (F-1)T_s]\left(\frac{S_o}{N_o}\right)_{min} \qquad (9.9)$$

This is the important final expression for the MDS. A practical example is given next.

Example 9.3

A microwave receiver has the following characteristics:

- Bandwidth 350 kHz;
- Noise figure 7 dB;
- Antenna noise temperature 800K;
- Minimum output signal-to-noise ratio 14 (linear factor).

What is the minimum detectable signal power level?
Substituting the above parameter values into (9.9) gives

$$S_{i_{min}} = (1.38 \times 10^{-23})(350 \times 10^3)[800 + (5-1)](14)$$
$$= 1.35 \times 10^{-13} \text{ W}$$

or, in dBm: −98.7 dBm, that is, approximately −100 dBm

For this receiver, the calculated MDS of almost −100 dBm represents a good low value for the minimum acceptable (detectable) input RF signal power level.

It is important to observe that the noise figure is always a significant parameter in determining the MDS in any receiver.

9.5 Noise in Transistors

Amplifiers, oscillators, and several other electronic circuits implement active devices (discrete transistors, MMICs, and RFICs). Apart from thermal noise, every transistor also introduces several further important sources of noise, regardless of whether the transistor is a discrete device or is within an IC. Developed from the FET-type equivalent circuit (Chapter 2), the important noise sources can be inserted, resulting in the noise-inclusive equivalent circuit of Figure 9.4.

The following four noise sources can be identified in Figure 9.4:

1. A thermal noise voltage source (V_{ng}) originating from the gate resistance;

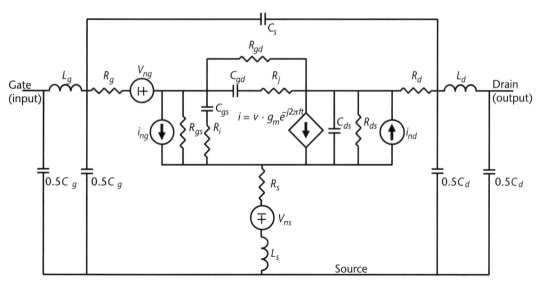

Figure 9.4 FET-type equivalent circuit, including noise sources. (© Artech House, 2016.)

2. A thermal noise voltage source (V_{ns}) originating from the source resistance;
3. A shot noise current source (i_{ng}) originating at the gate;
4. A shot noise current source (i_{nd}) originating at the drain.

Expressions for thermal noise and shot noise are given in following sections. However, it will always be necessary in practice to adjust well-calibrated, measured results to fit actual performance characteristics.

As mentioned in Chapter 2, the noise generated in a FET can be significantly decreased by the alloying-in of a halogen (e.g., fluorine, chlorine, bromine, iodine, or astatine) into the gate dielectric [4]. The all-important noise sources are defined and discussed in the following sections.

9.5.1 Thermal Noise, Particularly Thermal Diffusion Noise

All resistive and similar elements exhibit thermal noise. The noise power is given by (9.1), which is repeated here:

$$N = kTB$$

For a resistive element, value in (Ω), the RMS noise voltage generated is

$$v_n = \sqrt{4kTBR} \qquad (9.10)$$

where all the remaining parameters have been defined in Sections 9.1 and 9.2.

In a transistor thermal diffusion, noise is generated as a result of random variations in carrier speeds in the channel of a FET (e.g., a GaAs FET or a GaAs pHEMT). Extensive further information has been provided by Ladbrooke [5].

9.5.2 Shot Noise

Shot noise arises from junction-type semiconductor devices such as diodes, and BJTs. Equation (2.7) of Chapter 2 expresses both shot and also flicker noise for a semiconductor diode. The first term on the right side of (2.7) (which is the Schottky formula) can be reexpressed more generally as [6]:

$$I_{shot} = \sqrt{2qBI_{DC}} \tag{9.11}$$

where q is the charge on an electron (1.602×10^{-19} C), B is the noise bandwidth (Hz), and I_{DC} is the DC current flowing through the junction.

Because shot noise is independent of the frequency, it can be classified as white noise. Shot noise tends to be negligible in most instances, unless the noise bandwidth is exceptionally wide.

9.5.3 Flicker Noise

This type of noise is generated when a DC current flows through (typically) the interface between two conductive materials. Flicker noise (sometimes called $1/f$ noise) arises fundamentally from interface imperfections. Again, as mentioned above, (2.7) expresses both shot and also flicker noise for a semiconductor diode. However, the expression given in (2.7) for flicker noise is somewhat oversimplified and a more complete expression for flicker noise current is [7]

$$I_{flicker} = \sqrt{\frac{K_f I_{DC}^m B}{f^n}} \tag{9.12}$$

in which K_f is the flicker noise coefficient (dimensions amperes), I_{DC} is the DC flowing through the interface (amperes), B is the noise bandwidth, and f is the frequency (both in hertz) m is the flicker noise exponent, generally $1 \leq m \leq 3$, and n is the frequency exponent, which is mostly approximately 1. The general expression for K_f is [7]:

$$K_f = 5.333kTf_{corner}\sqrt{\frac{I_{DSS}}{V_{po}^2 I_D}} \tag{9.13}$$

in which k and T have already been defined and the other quantities are:

- f_{corner} = the flicker noise corner frequency, the frequency at which the noise spectral density is 3 dB higher than the voltage at high frequency;
- I_{DSS} = the saturated drain current (when $V_{GS} = 0$);
- V_{po} = the pinch-off voltage (usually between 3V and 4V);
- I_D = the drain current (A).

Example 9.4

Calculate the flicker noise current for an RF FET under the following parameters: f_{corner} = 30 MHz, I_{DSS} = 4 mA, V_{po} = 3.5V, I_D = 3 mA, flicker noise exponent = 2, T = 300K, and noise bandwidth = 350 kHz.

First, using (9.13), compute the value of K_f:

$$K_f = 5.333 \times 1.38 \times 10^{-23} \times 300 \times 3 \times 10^7 \times \sqrt{\frac{4}{3.5^2 \times 3}}$$

(noting that mA cancel under the radical). Hence,

$$K_f = 2.179 \times 10^{-13} A$$

Substitute this result together with the remaining data into (9.12):

$$I_{flicker} = \sqrt{\frac{2.179 \times 10^{-13} \times 9 \times 10^{-6} \, 3.5 \times 10^5}{3 \times 10^8}}$$

From which calculation it is only necessary to examine the orders of magnitude to see that the result is of the order 10^{-24} A. This is an extremely small current. In general, flicker noise on its own can be neglected, unless for some reason the noise bandwidth is exceptionally high.

However, flicker noise enters into the expressions for phase noise, which is very important and is the subject of the next section.

9.5.4 Phase Noise

Every amplifier and also indeed every frequency source suffer a further (highly distinct) form of noise called phase noise. This is an extremely important phenomenon that adversely affects the performance of LNAs as well as frequency sources including oscillators and frequency synthesizers. It is easiest to first consider an oscillator's ideal output spectrum vis-à-vis a practical spectrum where phase noise and spurious tones are present. The spectra shown in Figure 9.5 illustrate this comparison.

When any spot frequency in the response of an amplifier is examined using a spectrum analyser, the picture will be broadly similar to that shown in Figure 9.5(b) and certainly not at all like Figure 9.5(a). Camarchia et al. [8] provided a particularly interesting and concise treatment of phase noise, and, although it is highly oscillator-oriented, this treatment clearly indicates that both thermal and flicker noise sources are inherently involved. Since this is also particularly relevant to frequency sources, it is presented in some further detail in Chapter 12.

It is essential to consider noise power levels across the full spectrum of an amplifier; this is outlined in Figure 9.6.

In Figure 9.6, the low-frequency flicker noise corresponds to that discussed in Section 9.5.3. It is the existence of the flicker noise around the carrier frequency f_0 that gives rise to the phase noise. This occurs because nonlinearity associated with

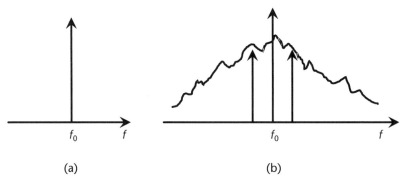

Figure 9.5 (a) Ideal (perfect) oscillator output spectrum and (b) actual (practical) output spectrum contaminated with phase noise and spurious tones.

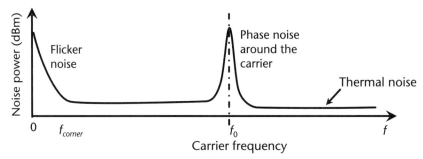

Figure 9.6 Full spectrum of noise power for a generalized amplifier.

the transistors within the amplifier modulates the carrier and hence produces $1/f$ noise sidebands situated symmetrically on each side of the carrier frequency. The process is directly analogous to amplitude modulation, which is covered in Chapter 14.

Therefore, only one side of the carrier needs to be studied from the phase-noise viewpoint and the upper-frequency side is usually chosen. The noise power spectral density amplifier for the LNA is K/f close-in, where K is a constant determined by the general level of the flicker noise. This is dominated by the modulating effect of the flicker noise and it produces the phase noise. As frequencies increase well above f_0, thermal noise (kT_sF/P_o) dominates.

Low-phase-noise LNAs represent an important subset within the broad scope of LNAs. They are particularly important in critical applications.

9.5.5 Variation of Noise Figure with Frequency

In general, noise factor (and therefore noise figure) deteriorates as operating frequency increases (i.e., at high frequencies, it is increasingly difficult to design LNAs with acceptably low noise figures). In this section the rationale behind this fact is developed. Also, some typical early and mid-twenty-first century state of the art noise figures are presented to a base of frequency.

In order to understand the reason why noise factor increases with frequency, it is necessary to return to (9.3) and also to Figure 9.4, the equivalent circuit for a FET (noise sources included). Equation (9.3) is

$$F = \frac{N_o}{GN_i}$$

Assume for this purpose that the ratio of the noise power terms remains approximately constant (i.e., invariant with frequency). However, the power gain of the LNA is a factor in the denominator of (9.3) and this power gain will decrease as a function of frequency due to the effects of the shunt capacitances and series inductances associated with the transistors (Figure 9.4).

Therefore, the noise factor as given by (9.3) must increase with frequency. This is exemplified by the data shown in Figure 9.7. ("Polynomial (F (dB))" refers to a best polynomial fit to the data.)

Some of the data presented in Figure 9.7 originated from [9, 10].

From Figure 9.7, it should be clear that LNAs with noise figures in the range of $0.5 \le F \le 1.6$ (dB) are regularly realized at the low-through-microwave frequencies (i.e., across the frequency range: $0.1 \le f \le 18$ GHz). Associated power gain values are generally around 20 dB.

Contrastingly, at higher frequencies (particularly through the millimeter-wave bands), best-value noise figures increase through 2 dB and onwards through 4.5 dB at 40 GHz. Over the first standard millimeter-wave 5G band (26–30 GHz), LNAs have noise figures around 2.5 to 3 dB. Overall receiver design must be able to accept and accommodate these relatively modest noise figure values.

LNAs designed using transistors based upon graphene can exhibit exceptional noise performance. However, there are immense stability issues and manufacturing challenges associated with this technology.

9.6 Overall Noise Figure for Cascaded Blocks

Practical receivers generally comprise a sequence of cascaded circuit blocks. Each block will have its own, separate noise figure and gain (or loss) and the vital ques-

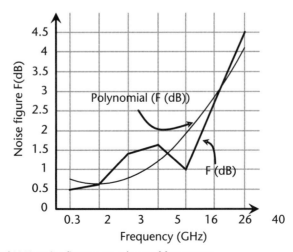

Figure 9.7 Selected LNA noise figures, to a base of frequency.

tion is: What is the overall noise figure? Two such cascaded circuit blocks are shown in Figure 9.8.

In Figure 9.8:

- F_1, G_1, T_{e1} are the noise factor, power gain and equivalent noise temperature of the particular block.
- Square boxes are shown because these blocks may or may not be amplifiers.

In the analysis that follows, the approach due to Pozar [2] is broadly followed. Pozar's approach is based on equivalent noise temperatures, which is a particularly effective and reliable method.

Based on (9.1), the input noise power N_i is given by

$$N_i = kT_s B$$

Adopting equivalent noise temperatures (often the safest approach), the output noise power from block 1 is

$$N_1 = G_1 k T_s B + G_1 k T_{e1} B \tag{9.14}$$

and the output noise power emerging from the second stage is

$$N_o = G_2 N_1 + G_2 k T_{e2} B$$

Substituting (9.14) into this expression and taking the common factor:

$$N_o = G_1 G_2 k B \left(T_s + T_{e1} + \frac{T_{e2}}{G_1} \right) \tag{9.15}$$

However, for the overall system comprising the two cascaded blocks:

$$N_o = G_1 G_2 k B (T_e + T_s) \tag{9.16}$$

Equating the right sides of (9.15) and (9.16) and solving for T_e gives the important result:

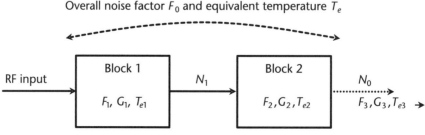

Figure 9.8 Cascaded circuit blocks.

9.6 Overall Noise Figure for Cascaded Blocks

$$T_e = T_{e1} + \frac{T_{e2}}{G_1} \qquad (9.17)$$

Expressing this result in words, the overall equivalent noise temperature equals the equivalent noise temperature of the first stage taken alone, plus the equivalent noise temperature of the second stage taken alone, but divided by the power gain of the first stage (G_1).

Now, appropriately substituting (9.4) into (9.17) yields:

$$(F-1)T_s = (F_1 - 1)T_s + \frac{(F_2 - 1)T_s}{G_1}$$

Dividing through by T_s and adding 1 to each side of this expression yields the final result for the overall noise factor of two cascaded blocks:

$$F = F_1 + \frac{(F_2 - 1)}{G_1} \qquad (9.18)$$

Expressing this result in words, the overall noise factor equals the noise factor of the first stage taken alone, plus the noise factor of the second stage taken alone (with 1 subtracted) and divided by the power gain of the first stage (G_1).

By observing how in each successive term the temperature is divided by the total power gain of the preceding blocks (or stages), this important result, (9.18), can be generalized to yield the overall noise factor applying to any number of cascaded stages (again starting with the equivalent noise temperature and power gains for three cascaded blocks):

$$T_e = T_{e1} + \frac{T_{e2}}{G_1} + \frac{T_{e3}}{G_1 G_2} + \ldots \qquad (9.19)$$

and reconverting back to noise factors this expression becomes:

$$F = F_1 + \frac{(F_2 - 1)}{G_1} + \frac{(F_3 - 1)}{G_1 G_2} \qquad (9.20)$$

Clearly, the first stage or block is the most critical in terms of the effects of noise in the receiver chain because its noise effect (factor or temperature) stands alone, that is, for the first stage is not reduced (by 1 as with all succeeding stages) nor is it divided by preceding power gain factors.

Example 9.5

The first three stages of a receiver subsystem comprise amplifiers with the following parameters:

G = 8 dB   G = 12 dB   G = 20 dB
F = 1.4 dB F = 3.4 dB   F = 10 dB

What is the overall noise figure for this subsystem?

First, it is vital to appreciate that all the above expressions are for noise factors, and they all involve noise factors (and power gain factors) within their terms. Therefore, it is essential to first convert all the quantities in decibels to (linear) noise factors and gain factors:

- Noise figures: $F = 1.4$ dB, $F = 3.4$ dB, $F = 10$ dB
- Noise factors: $F = 1.38$, $F = 2.19$, $F = 10$
- Gain values (dB): $G = 8$ dB, $G = 12$ dB, $G = 20$ dB
- Gain values (linear): $G = 6.31$, $G = 15.85$, $G = 100$

Substituting the appropriate values into (9.20):

$$F = F_1 + \frac{(F_2 - 1)}{G_1} + \frac{(F_3 - 1)}{G_1 G_2}$$

$$F = 1.38 + \frac{2.19 - 1}{6.31} + \frac{10 - 1}{6.31 \times 15.85}$$

$F = 1.66$ (which is just 20% higher than the first stage value of 1.38)
In decibels (for the noise figure), this is $F = 10\log 1.66$ or $F = 2.2$ dB.

Example 9.6

The first three stages of a receiver subsystem comprise circuits having parameters as indicated in Figure 9.9. (Compare and contrast with Example 9.5.)

What is the overall noise figure for this subsystem?

For this subsystem, the bandpass filter (BPF) must be treated exactly as for the attenuator (Section 9.3), where (9.6) is repeated here:

$$F = 1 + (L - 1) T/T_s$$

Also, it will be assumed that $T = T_s$.

Applying this calculation, the BPF noise factor is 1.148 and the mixer loss factor is 1.585. To find the BPF noise factor, it is necessary to use (9.6). The result is:

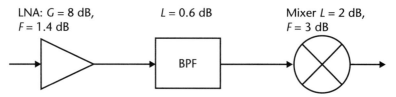

Figure 9.9 Trio of cascaded (matched) circuits.

9.6 Overall Noise Figure for Cascaded Blocks

$$F = 1 + (1.148 - 1) = 1.148$$

The mixer is treated like an amplifier but with a loss (i.e., −2 dB) instead of a gain, which leads to a loss factor of 0.631.

Then the values become:

- Noise figures: $F = 1.4$ dB, (BPF noise factor already known), $F = 3$ dB
- Noise factors: $F = 1.38$, $F = 1.148$, $F = 1.995$
- Gain (or loss) values (dB): $G = 8$ dB, $L = 0.6$ dB, $L = -2$ dB
- Gain values (linear): $G = 6.31$, $L = 0.871$, $L = 0.631$

Then again substituting the appropriate values into (9.20):

$$F = F_1 + \frac{(F_2 - 1)}{G_1} + \frac{(F_3 - 1)}{G_1 G_2}$$

$$F = 1.38 + \frac{1.148 - 1}{6.31} + \frac{1.995 - 1}{6.31 \times 0.871}$$

which gives the final result for the noise factor $F = 1.584$.

In decibels (for noise figure), this is $F = 10\log 1.584$ or $F = 1.9$ dB.

Comparing with Example 9.5, this result is lower (i.e., better) by 0.3 dB. This shows that even where lossy elements (the BPF and mixer stages) are present, the noise figure can often still be better than the subsystem with further amplifiers (Example 9.5). The substantial noise figures of these further amplifiers frequently represent a major consideration in the design of the overall subsystem.

In the preceding analysis and examples, it has been assumed that all the blocks are input-matched. When this is not the case (e.g., with some attenuators or mixers), an appropriate extra coefficient must be included in the final expression. The effect of this mismatch must then be included in the expression for the overall noise factor for the cascaded blocks or stages. Without proof, where the mismatch coefficients are M_1, M_2, M_3, ..., M_n (equivalent to the voltage standing-wave ratios [VSWR]) for each amplifier, then (9.20) becomes:

$$F = F_1 + \frac{(F_2 - 1)M_2}{M_1 G_1} + \frac{(F_3 - 1)M_3}{M_1 M_2 G_1 G_2} \qquad (9.21)$$

In practical terms, it is always important to remember the following:

- Before using the noise factor expressions each parameter must be converted from decibels to linear factors.
- From fundamentals, it is always best to start with equivalent noise temperatures because these lead to simpler, more direct expressions. Then (9.4) can be implemented to convert to noise factors.

9.7 Noise-Matching and Narrowband LNA Design

Designing an LNA requires several steps including transistor choice, selection of the appropriate DC bias, and passive network synthesis. In this section, these steps are considered, including a detailed example of the required passive network synthesis.

The three major steps required start with transistor specifications that are usually available from the transistor manufacturer's Websites. This treatment will assume that a discrete transistor is being selected, whereas in most real-world examples the transistor will be based on the MMIC/RFIC process that has been chosen for the LNA design. The steps are:

- Select an appropriate transistor for the low noise application (probably a GaAs pHEMT or maybe a SiGe HBT for some applications).
- Select an appropriate DC bias (I_c or I_d) for the transistor such that the minimum noise figure can be obtained from the device (this should be provided by the device manufacturer, but in many instances careful measurements by the system designer should be followed through to check).
- Design a noise-matching (or noise-tuning) network for the device. This will need to be interposed between the input (normally the gate) and the signal source, the source generator.

From this point onward in this section, it is assumed that the first two requirements have been met and the third requirement of noise-matching is to be performed.

The general principle of noise-matching is indicated in Figure 9.10.

The aim is to minimize the noise figure of the overall arrangement from the source port through to the transistor output, that is, to establish F_{min}. Toward this, the passive noise-matching or noise-tuning network must be designed. Two critical impedances must first be defined:

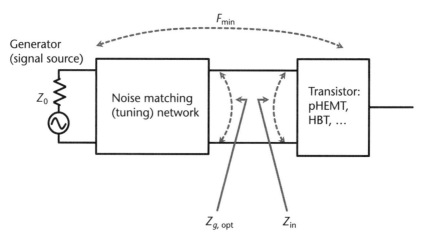

Figure 9.10 Schematic diagram showing the general principle of noise-matching.

9.7 Noise-Matching and Narrowband LNA Design

- $Z_{g,opt}$ is the driving-point impedance working backwards towards the source generator (i.e., leftwards in Figure 9.10). It is to this value that the design of the noise-matching network must be established.
- Z_{in} is the transistor's input impedance.

This impedance may be provided in the manufacturer's specification or can be determined from the device's characteristics (e.g., by measurement). In principle, for any given amplifier, there exists an optimum source (or generator) impedance that will yield F_{min}.

The challenge is to design the required noise-matching passive network so as to synthesize $Z_{g,opt}$. For narrowband applications a rotated-L-shaped transmission-line (microstrip) configuration often suffices and a typical structure is shown in Figure 9.11.

In this arrangement, Z_{0a} is the characteristic impedance of the first, horizontal, microstrip, and Z_{0b} is the characteristic impedance of the second, vertical, microstrip.

The corner chamfer minimizes the electrical discontinuity effect. The guide wavelength (λ_g) is the wavelength of the approximately TEM wave in the microstrips (actually somewhat different in each line section). All the required microstrip design approaches and expressions are provided in Chapter 3. For practical LNA MMIC design, any of the available comprehensive EDA (CAD and simulation) packages can be used (e.g., Agilent Technologies' ADS, NI AWR).

Referring to the microstrip layout indicated in Figure 9.12, the quarter-wave transformer, $\lambda_g/4$ in length, has characteristic impedance Z_{0a}, and the eighth-wavelength section of line, $\lambda_g/8$ in length, has characteristic impedance Z_{0b}. This section of line represents an inductive reactance having input value jZ_{0b}.

In detail, the three design (synthesis) steps are therefore:

1. Determine $Z_{g,opt}$, convert this value to an equivalent reflection coefficient Γ_0, and thence to an admittance that is normalized to the characteristic admittance, termed Y_{NF}.
2. Determine the $\lambda_g/8$ line parameters so as to realize the susceptive part of Y_{NF}, which is then converted to the equivalent reactance.

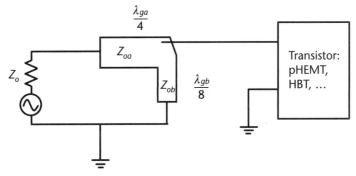

Figure 9.11 Conceptual layout of a single-stage narrowband LNA.

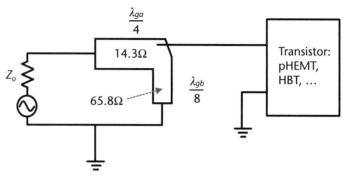

Figure 9.12 Layout of the single-stage narrowband LNA of Example 9.7.

3. Design the $\lambda_g/4$ transforming line section so as to match the (now entirely real) impedance (looking backwards towards the generator from the corner of the two sections in Figure 9.12) to the generator impedance Z_0.

Example 9.7

Toward an LNA design, it is found that:

1. The real part of $Z_{g,opt}$ is 42Ω.
2. The reflection coefficient Γ_0 is 0.7 ∠ 20°.

Also assume that Step 1 has been completed, determining that the input reactance of the $\lambda_g/8$ element equals $j65.8Ω$.

The $\lambda_g/4$ transforming line section must have characteristic impedance Z_{0a} and this is determined using a fundamental expression from transmission-line theory, in terms of Γ_0:

$$Re(Z_{g,opt}) = \frac{(1-|\Gamma_0|^2)Z_{0a}}{1+|\Gamma_0|^2 - 2|\Gamma_0|\cos\varphi_0} \tag{9.22}$$

From Step 1, the left side of this expression equals 42Ω and the only unknown quantity of the right side is Z_{0a}. Equation (9.22) can therefore be directly solved for Z_{0a}. The resulting value is: $Z_{0a} = 14.3Ω$.

The characteristic impedance of the $\lambda_g/8$ line is already known (it is the magnitude of the reactance, which is already given above). Therefore, the final network parameters are known, except for the physical line lengths, which can easily be calculated from expressions given in Chapter 3. The final network is shown in Figure 9.12.

Some interesting and useful practical guidelines can be drawn from these results. Summarizing, the impedance looking backwards towards the generator from the intersection point between the two line sections is:

$$Z_{g,opt} = 42.0 + j65.8Ω$$

In general, for GaAs pHEMTs and HBTs (and MMICs implementing these processes), it can be expected that $Z_{g,opt}$ will be in the region of the following broad values:

$$Z_{g,opt} \sim R + jX$$

in which $|R|$ and $|X|$ are both of the order of some tens of ohms, which turns out to be a good general rule. However, it should be appreciated that device (chip) process variations usually render $Z_{g,opt}$ highly variable even within production batches. Hence, it becomes very problematic to expect a synthesis of a $Z_{g,opt}$ network to be at all accurate across a batch, in particular.

Having said this, even with a nominal design the resulting actual noise figure is usually reasonably close to F_{min} even when $Z_{g,opt}$ is way off the theoretical value by, say, $\Delta Z_{g,opt}$. This situation can be summarized as:

with the important proviso that $\dfrac{\Delta Z_{g,opt}}{Z_{g,opt}} \sim 20\%$ or 30%

Then the actual noise figure will be close to the ideal F_{min}.

The effects of transistor noise as well as noise matching can be summarized. To design an LNA with minimum possible noise figure, the device selection and the bias current are critical. However, it is also necessary to insert an appropriate noise-matching or noise-tuning network between generator and device input to optimize the entire circuit for truly minimum noise.

The above procedure relates specifically to narrowband design. For broader-band design, one approach is to implement several, cascaded, rotated-L-shaped configurations of this general type, but this is a particularly clumsy method that eats up die area in a MMIC. Broadband design is sufficiently complex to demand EDA.

Alternative broadband amplifier approaches are the best option for broadband, low-noise characteristics. Some distributed amplifiers, for example (Chapter 8) can operate over the extremely wide bandwidth of DC to 50 GHz with noise figures around 3.5 dB across the band.

References

[1] Browne, J., "Trying to Keep the Noise Down," *Microwaves & RF*, August 2017, pp. 39–40.

[2] Pozar, D. M., *Microwave and RF Design of Wireless Systems*, New York: John Wiley & Sons, 2001.

[3] Kingsley, N., and J. R. Guerci, *Radar RF Circuit Design*, Norwood, MA: Artech House, 2016.

[4] https://patents.google.com/patent/US8076228B2/en.

[5] Ladbrooke, P. H., *MMIC Design: GaAs FETs and HEMTs*, Norwood, MA: Artech House, 1989.

[6] Ott, H., *Noise Reduction Techniques in Electronic Systems*, New York: John Wiley & Sons, 1988.

[7] Leach, W., "Fundamentals of Low-Noise Analog Circuit Design," *Proceedings of the IEEE*, Vol. 82, No. 10, October 1994.

[8] Camarchia, V., R. Quaglia, and M. Pirola, *Electronics for Microwave Backhaul*, Norwood, MA: Artech House, 2016.

[9] Lim, C. -L., "Low-Noise Amplifier Aids TDD Small Cells," *Microwaves & RF*, July 2017, pp. 44–48 and 82.

[10] Shinjo, S., et al., "Integrating the Front End," *IEEE Microwave Magazine*, July/August 2017, pp. 31–40.

CHAPTER 10

RF Power Amplifiers

10.1 Introduction

Every transmitter or transceiver requires a final RF power amplifier (RFPA) to boost the power level prior to delivery to the antenna. There are many constraints applying to the design of RFPAs and the purpose of this chapter is to provide substantial details concerning such constraints. Fundamental concerns include the choice of underlying technology, notably whether to go for a hybrid design implementing discrete transistors or alternatively whether it is appropriate to choose an integrated design approach (i.e., MMIC). Following the initial choice of technology (mainly the type of transistor), then CAD and computer-based simulations using electronic design automation (EDA) are extensively used for designing and simulating RFPAs.

Many references provide substantial information regarding RFPAs, in various contexts. Good examples include the books by Cripps [1], Camarchia et al. [2], and Kingsley and Guerci [3].

Several aspects presented in Chapter 8 also apply in the context of RFPAs.

10.2 Some Basic Aspects of RFPAs

The block triangle shown in Figure 8.1 is a basic symbol generally used to represent any amplifier, therefore including any RFPA.

This is a type of signal flow representation in which other important aspects such as common earth and DC power supplies are omitted for clarity. The symbol G refers to the power gain of the amplifier (i.e., the signal power P_o delivered from the output divided by the signal power delivered into the input P_i as indicated in [10.1]):

$$G = P_o / P_i \tag{10.1}$$

When P_o and P_i are expressed in watts or milliwatts, then clearly from (10.1) G is a numeric ratio (i.e., a factor such as 4.3, 27, or 50). However, G is often expressed in decibels rather than as a numeric. Also the input and output power levels are frequently expressed in dBm or occasionally in dBW.

Although amplifier power gain can be defined in various alternative ways (e.g., transduced power gain) for all purposes throughout this book, the relatively simple description given by (10.1) will suffice.

Most of the specification aspects applicable to an RF power amplifier are cited in Chapter 8 for general amplifiers and the great majority of those aspects apply here.

10.3 Transistor Choices, Hybrid Circuits, and MMICs

The basic requirement for any solid-state amplifier is the internal transistor and these fundamental semiconductor devices are described in Chapter 2.

While the main thrust of technology choice is toward MMIC realizations, discrete transistors are required where there is a need for relatively high output power, generally upwards of several tens of watts (CW) or kilowatts (pulsed), and/or scenarios where the RF output power may be relatively low (generally below a few tens of watts), but custom or low production rate designs are the order of the day.

As an example, a 100-W RFPA operating around 2 GHz typically implements one or more discrete GaN HEMTs and such a design would almost certainly take the hybrid circuit route, most likely on a polymer-based circuit board with excellent heat-sinking.

In contrast, an RFPA required to provide a 10-W output at moderate frequencies would very likely be designed in MMIC format, provided that the production rate is at least several thousand pieces. The transistor process in this case will depend mainly on the signal frequencies involved: typically GaN HEMTs for lower microwave frequencies, although more likely GaAs pHEMTs for designs around or above 28 GHz.

However, the strong trend is toward silicon transistor processes for the lower-power scenarios. All these types of processes are described in Chapter 2.

As hinted above, the selection of the transistor process depends critically on the operating frequency. This feature is a consequence of the internal and parasitic reactive elements associated with every transistor, which again are described in Chapter 2.

These reactive effects strongly influence the RF output power and power gain as functions of frequency for any transistor together with RFPAs implementing these devices. Considering microwave power transistors and RFPAs using them, the values and trend shown in Figure 10.1 apply.

The values indicated in Figure 10.1 clarify the fact that while a 100-W-plus circuit is feasible at 1 or 2 GHz, where frequencies reach 20 GHz an 8-W amplifier is more realistic. For frequencies up to around 10 or 12 GHz, discrete GaN HEMTs will be implemented into hybrid circuitry. At higher frequencies, with lower power levels, MMIC design is more appropriate.

Under millimeter-wave conditions, RF output power levels still decrease with frequency and the choice of process technology changes radically, as shown in Figures 10.2 and 10.3.

In Figures 10.2 and 10.3, it is important to observe that the RF power is in milliwatts, compared with the power in watts indicated in Figure 10.1. As a result, the great majority of designs at these frequencies tend to be MMIC-based.

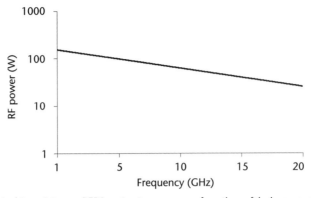

Figure 10.1 Typical transistor or RFPA output power as a function of (microwave) frequency.

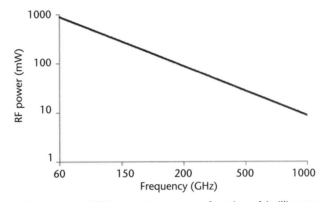

Figure 10.2 Typical transistor or RFPA output power as a function of (millimeter-wave) frequency.

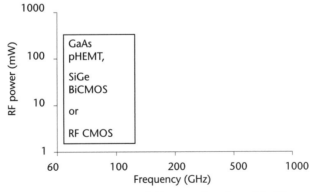

Figure 10.3 Typical transistor processes available, as a function of (millimeter-wave) frequency (millimeter-wave-level SiGe BiCMOS or RF CMOS above 150 GHz).

10.4 Power Levels, Power Gains, and Efficiency

10.4.1 Internal Transistor Output Characteristics

Throughout this chapter, necessarily, reference will frequently be made to power amplifier output versus input power characteristics. These characteristics are always highly nonlinear and the rationale behind this fact is explored here.

The typical DC I_D/V_D characteristics for any FET-type transistor are shown in Figure 2.15 of Chapter 2.

This important graph is returned to in support of descriptions of amplifier classifications later in this chapter.

In general terms, these characteristics will apply to GaN HEMTs, GaAs pHEMTs, each device in a CMOS circuit, and so on. The major difference is in the scaling, for example:

- I_d in mA and V_d up to around 10V or 12V maximum for a GaAs pHEMT;
- I_d in amperes and up to around 40V or 50V for a GaN HEMT.

Returning to Figure 2.15 it is also clear that the curves are particularly nonlinear, with rapidly increasing drain current at low drain voltages before turning to quasi-saturate as the drain voltage increases further. Since these transistors form the backbone of any RFPA, it is evident that they will, in turn, cause similarly pronounced nonlinearity in the final RFPA circuit.

10.4.2 RFPA Output-Input Power Transfer Characteristics

As a result of transistor nonlinearity as described above, the basic output-input power transfer characteristic for almost any PA will look like the curve indicated in Figure 1.20.

It can be seen in Figure 1.20 that the output power P_o increases almost linearly with increases in the input power P_i until this input power reaches higher levels, where increases in the output power begin to slow down. From the point P'_o, P'_i (where P'_o is the corresponding backed-off input power), onward further increases in the input power progressively result in ever-smaller increases in the output power until eventually a maximum value of P_o is reached where what is termed saturation sets in (P_{osat}, P_{isat}). All power levels are almost always expressed in dBm.

So as to remain well removed from this saturation region, most RFPAs are operated in backed-off states, that is, they are designed to ensure operation that maximizes at the P'_o, P'_i point. As a representative example in a specific RFPA, P_{osat} could be 32 dBm, with the corresponding $P_{isat} = 20$ dBm. Back-off values could be 4 dB and 7 dB, respectively, yielding (after back-off) $P'_o, P'_i = 28$ dBm and 13 dBm for the maximum useable power levels in this instance.

10.4.3 Amplifier Efficiency

It is impossible for any practical amplifier to be able to convert all of the supplied DC power into output RF power. Inevitably some of the supplied DC power will become lost as heat and for a medium-to-high power amplifier a heat sink will be needed to ensure the circuit module will not overheat, which would otherwise result in degraded performance or even device destruction.

The ability of any amplifier to convert some of the DC supplied power into RF output power is termed the basic efficiency of the amplifier circuit. The simplest

10.4 Power Levels, Power Gains, and Efficiency

definition of basic efficiency is (RF output power)/(DC supplied input power) or, more formally:

$$\eta = \frac{P_o}{P_{DC}} \times 100(\%) \tag{10.2}$$

where the amplifying device is any type of FET (including GaN HEMT, GaAs pHEMT), then (10.2) is often referred to as the drain efficiency (η_{drain}).

This definition of efficiency works fairly well so long as the input signal power is relatively small, (i.e., $P_i \ll P_o$). In most practical instances, however, this is not the case and therefore P_i must be subtracted from P_o in the numerator of (10.2), yielding what is termed power-added efficiency (PAE, or η_{PAE}). The expression for this is therefore

$$\eta_{PAE} = \frac{P_o - P_i}{P_{DC}} \times 100(\%) \tag{10.3}$$

PAE is a much more realistic measure of the efficiency of an RFPA.

As a practical example assume the same amplifier as specified in Section 10.4.2, operating with its backed-off power levels (i.e., 28 and 13 dBm for the output RF power and the input power, respectively). The aim is to find the PAE where this amplifier is operating with a 30-V DC supply rail and is drawing 50-mA of DC (drain) current.

First, the power levels (specified or measured in dBm) must be converted to linear values, giving 631 mW and 20 mW for the output and input signal power levels. Using all these values and substituting into (10.3), using milliwatt and milliampere quantities gives

$$\eta_{PAE} = \frac{631 - 20}{30 \times 50} \times 100(\%)$$

Leading to: $\eta_{PAE} \approx 40.7\%$
which is fairly representative of the order of value for PAEs in RFPAs.

Observe that use of the basic efficiency expression, (10.2), yields 42.1% which is slightly optimistic comparing η_{PAE}.

As a practical point, it is emphasized that manufacturer's (or other suppliers) quotations of amplifier efficiency must be carefully inspected to discern whether they are referring to η_{PAE} or to the basic efficiency that neglects the input signal power.

Most RF and microwave amplifiers have η (and PAE) well below 50% because they mostly operate in Class A to take advantage of the high linearity. (Power amplifier classifications are considered later in this chapter.)

PAE can also be expressed as a function of basic efficiency and the power gain by combining (10.1) and (10.3), to yield

$$\eta_{PAE} = \eta\left(1 - \frac{1}{G_p}\right) \times 100 (\%) \qquad (10.4)$$

Therefore, knowing the basic efficiency and the power gain enables the PAE to be calculated immediately.

10.5 Compression and Peak-to-Average Power Ratio

10.5.1 Compression and a Summary of Main Parameters

An important distinction must be made between small-signal and large-signal (or effective) power gains. Small-signal power gain G_{ss} is associated with relatively high linearity as well as bias-point dependence. G_{ss} can be determined using the small-signal scattering ([S]) parameters. However, large-signal power gain G_p will be associated with at least a certain amount of distortion due to the inherent nonlinearities already discussed. G_p cannot be determined using the [S] parameters but instead requires special parameters or graphically derived quantities.

The difference between G_p and G_{ss} is termed the gain compression level. For example, if for some RFPA G_{ss} is 18 dB and G_p is 16 dB, then the gain compression is the difference (i.e., 2 dB).

Camarchia et al. [2] have published a particularly useful and comprehensive graph showing G_p, P_{out} and both efficiencies to a base of P_{in}. This graph is reproduced here, as Figure 10.4.

Points to observe considering the curves in Figure 10.4 include:

- G_{ss} and G_p both decrease very slowly until the amplifier approaches saturation, after which further increases in P_{in} cause power gain to begin decreasing relatively rapidly.
- The P_{out} versus P_{in} curve is broadly similar to that shown in Figure 1.20, except for the eventual downturn following saturation (Figure 1.20).

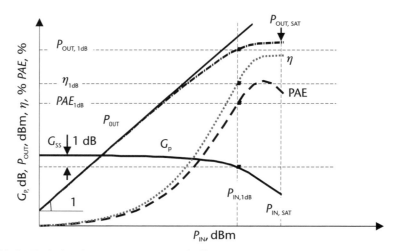

Figure 10.4 Typical gain, output power and efficiencies plotted to a base of input power [2]. (© Artech House, 2016.)

- The basic efficiency η steadily increases until flattening-out with the onset of saturation, however.
- PAE (also steadily increasing at first) then peaks just as saturation becomes evident, before declining with further increases in P_{in}.
- At low-to-moderate levels of P_{in}, there is 1 dB of compression between G_{ss} and G_p.
- A straight line is drawn, asymptotic to the P_{out} versus P_{in} curve at low levels. The importance of this will be clear when intermodulation is discussed later in this chapter.

10.5.2 Peak-to-Average Power Ratio

Most communications systems operate with variable-envelope modulation schemes in order to ensure good spectral efficiency (Chapter 14). Because of this scenario power levels must be treated statistically in terms of their probability density functions (PDFs). This feature impacts how the average output power must be computed.

For any power amplifier, the peak-to-average power ratio (PAPR) is defined by:

$$PAPR_{dB} = 10 \log_{10} \left(\frac{P_{out,peak}}{P_{out,avg}} \right) \tag{10.5}$$

where $P_{out,peak}$ is the maximum envelope output power anticipated (or measured), and $P_{out,avg}$ is the statistically computed average value given by:

$$P_{out,avg} = \frac{1}{T} \int_0^T P_{out}(t) PDF \, dt \tag{10.6}$$

according to which the output power is averaged across the envelope period T and then weighted by its PDF.

This scenario is particularly significant in mobile (cell phone) communications systems. A good example considers the standards applicable to LTE, backhaul and IEEE 802.11b. The impact of (10.5) and (10.6) shows that the required peak power output back-off decreases from 13 dBm to 6 dBm, progressing across these three standards. Camarchia et al. provided more details concerning this in their book [2].

10.6 Error Vector Magnitude

The definition of a quantity universally known as the error vector magnitude (EVM) provides a particularly important specification of system performance under complex modulation schemes. This performance is a pronounced aspect of RFPAs and the EVM indicates the difference between ideal constellation signals and those actually measured at the amplifier output.

EVM is defined as shown in Figure 10.5, in which I and Q are, respectively, the in-phase and quadrature modulation signal components.

It can readily be seen from Figure 10.5 that the magnitude of the error is the difference between the measured vector amplitude and that of the ideal vector. If, for example, the ideal vector has a magnitude of 3V and the measured vector has a magnitude of 3.2V, then the EVM is 0.2V or about 6.7%.

10.7 Classifications of Power Amplifiers

Power amplifiers are classified according to the manner in which the circuits are biased, and the resultant output signal swing. These features have profound implications regarding, in particular, linearity, efficiency, and gain. This is standard large-signal amplifier terminology and analysis (i.e., it is quite general and does not have to apply to RF circuits as such).

The main classifications A, B, AB, and C, are often termed transconductance amplifiers because of their reliance on the g_m of the transistors.

10.7.1 Class A Amplifiers

In relative terms, this is the most linear of all the classes of amplifier. Drain current versus gate voltage and FET output characteristics for a Class A power amplifier are shown in Figure 10.6.

The basic FET output characteristics are shown in Figure 2.15, shown earlier in this chapter. The important additional information shown in Figure 10.6 centers around the quiescent point labeled Q, having coordinates V_{DQ}, I_{DQ}. This quiescent or DC operating point always determines the class of amplifier, in this case with Q fairly central in the output characteristics it signifies a Class A amplifier.

Also within the output characteristics a load line is constructed. This line extends from the lowest current level ($I_{D,min}$) up to the highest available level (I_{knee}). Hence, this load line also dictates the maximum and minimum voltage and current swings (V_{swing} and I_{swing} in Figure 10.6) and the controlling gate voltage (V_g) swing as indicated on the left-side curve in Figure 10.6.

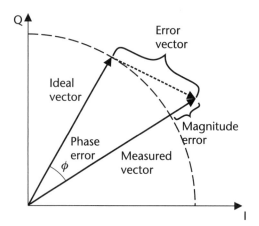

Figure 10.5 Phasor diagram illustrating EVM [2]. (© Artech House, 2016.)

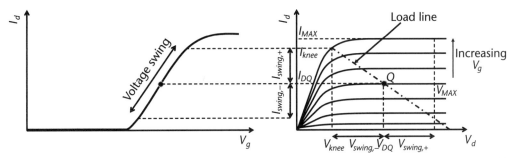

Figure 10.6 Drain current versus gate voltage and FET output characteristics for a Class A power amplifier [3]. (© Artech House, 2016.)

In terms of these voltage and current swings, the maximum possible output power can be expressed as

$$P_{max} = \frac{V_{pp}I_{pp}}{8} = \frac{(V_{swing,-}+V_{swing,+})(I_{swing,-}+I_{swing,+})}{8} \quad (10.7)$$

In which 8 in the denominator arises from the fact that power is the product of the two rms quantities ($P = V_{rms}I_{rms}$), $V_{swing} = \sqrt{2}V_{rms}$, $I_{swing} = \sqrt{2}I_{rms}$ and in magnitude terms each of the two swing extremes (voltage and current) amounts to a further doubling of the magnitudes.

Further observations regarding the Class A amplifier are:

- To preserve this class of operation, the absolute maximum input power level is $P_{max}/2$. The transistor is always conducting in this class of amplifier (i.e., the conduction angle is 360°).
- Since DC power is being delivered under the quiescent (approximate average) conditions, the theoretical maximum efficiency is 50%.

In practice however the voltage signals cannot swing to zero but instead are limited to the V_{knee} value (Figure 10.6). This feature reduces the attainable efficiency to well below 50% and it can be as low as 35%.

To achieve maximum output power the load impedance on a Class A amplifier should be set to:

$$Z_L = \frac{V_{pp}}{I_{pp}} \quad (10.8)$$

For example, if V_{pp} = 8V and I_{pp} = 0.5A, then Z_L should be made as close as possible to 16Ω (a relatively low impedance).

Most low-through-medium power microwave and millimeter-wave PAs operate in Class A mode.

10.7.2 Class B and AB Amplifiers

Camarchia et al. [1], in their Figure 6.9, have presented a particularly comprehensive diagram showing all classes A through C on the basis of the FET output characteristics. This is reproduced as Figure 10.7 here.

Class A, described fully in Section 10.7.1, appears at the top quiescent point. This is followed, downward, by Classes AB, B, and C, respectively. Class B is described next and as can be seen from Figure 10.7 the quiescent point for this class is set where the drain current is zero. This means that drain current is only drawn when the drain-source voltage is between the knee value and just approaching zero (at quiescent point B). When the drain-source voltage is higher than V_B, no drain current can flow.

Important voltages specified in Figure 10.7 are:

- $V_{DS,k}$ = the knee value of V_{DS};
- $V_{DS,br}$ = the breakdown (i.e., maximum) value of V_{DS};
- V_T = the value of V_{gs} corresponding to quiescent point B.

The main properties of a Class B amplifier are:

- Due to the highly nonlinear voltage swing, any Class B amplifier will be also be nonlinear in operation.
- The transistor in a Class B amplifier is only conducting for one-half of the full signal cycle (i.e., the conduction angle is 180°).
- Because the conduction is much less than that of a Class A amplifier, the power gain of a Class B amplifier must always be substantially lower than that of a Class A amplifier.
- The maximum theoretical efficiency is approximately 78.5%.

Under Class B operation, it is possible to implement two transistors, operating for the two half-cycles (i.e., a push-pull configuration).

However, although the maximum efficiency is high it is the poor linearity and gain characteristics that mitigate against the widespread adoption of Class B ampli-

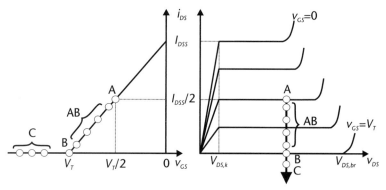

Figure 10.7 Drain current versus gate voltage and FET output characteristics for power amplifier classes A through C [2]. (© Artech House, 2016.)

fiers. The Class AB design approach is more attractive and four quiescent points relating to this class of operation are shown in Figure 10.7.

It can be seen that Class AB represents a compromise between Classes A and B and as a result properties always lie at some level between Class A and Class B:

- Linearity;
- Efficiency (i.e., theoretically between 50% and 78.5%);
- Power gain;
- Transistor conduction angle (between 180° and 360°).

Class AB power amplifiers are quite extensively adopted in practice, with a parallel tuned circuit connected across the output load impedance to prevent harmonic components from reaching the output. This technique is termed harmonic matching and further information on this is presented later in this section where Class E switched-mode RFPAs are considered.

10.7.3 Class C Amplifiers

Referring again to Figure 10.7, it is seen that biasing for Class C operation represents an extreme example of biasing well below pinch-off for the FET in this case. As a direct result of this approach, any Class C amplifier is the most power-efficient of all designs, but at the cost of easily the worst linearity.

Unlike any of the other classes of power amplifiers, in Class C only a relatively small positive current flow occurs. Class C amplifier properties are (comparing Class B):

- Linearity is worse;
- Efficiency is better (theoretically up to 81% maximum);
- Power gain is lower;
- Transistor conduction angle is smaller (typically between 90° and 120°).

The maximum theoretical efficiency is actually 100%, but this is when the output power is zero. For conduction angles within the range 90° to 180°, the efficiency is decreased to 81%.

10.8 Harmonically Matched Power Amplifiers

10.8.1 Switched-Mode RFPAs

10.8.1.1 Introduction to Switched-Mode RFPAs

In Section 10.7, it was pointed out that the performance of any Class B, Class AB, or Class C amplifier can be significantly improved by harmonic matching at the output. In practice, this means inserting a harmonic-absorbing tuned circuit between the transistor and the output load. It is possible to take this approach further, resulting in special classes of amplifiers such as Class E and Class F. In either case the aim is toward achieving as close as possible to an efficiency of 100%.

Although the various classes of power amplifiers covered in earlier sections of this chapter continue to apply to microwave and some millimeter-wave applications, it is often (misguidedly) considered that switched-mode power amplifiers are unsuitable for use at such high frequencies. The main reason cited has been the requirement for transistor switching times to be substantially shorter than even fractions of a period, which is already very short (e.g., the full period is 100 ps at 10 GHz).

The main class of switched-mode PAs, alternatively termed switching PAs, are harmonically matched Class E circuits which can have theoretical efficiencies around 100%. For Class E, the transistor is terminated with a high-Q resonant circuit that provides a reactive load at the desired center or fundamental frequency f_0 but also provides open circuits at the second and third harmonics $2f_0$ and $3f_0$. A theoretically exact implementation of this approach demands the instantaneous switching of the transistor into its on and off states. This switching process leads to the terminology switched-mode or switching power amplifiers.

Various professionals, especially those whose work is primarily concerned with microwave power amplifiers, notably Cripps and also Grebennikov [1, 4], have pointed out the error in believing that switched-mode PAs are inherently unsuitable for microwave applications. In the remainder of this section, examples are taken from Cripps's work, focusing on Class E switched-mode PAs. In all instances, Cripps provided extensive analyses, especially regarding calculations of efficiency and the related development of characteristic curves for the various power amplifier configurations. The starting point is the greatly simplified basic concept, which is shown in Figure 10.8.

This circuit is supplied with DC power, delivering current I_{dc} from the power supply rail at voltage V_{dc}. RF energy is blocked from returning to the power supply by means of the choke inductor.

The key element in this circuit is the switch which rapidly switches on and off, sustaining a voltage V_{sw} when off and delivering a current i_{sw} when in the on state. Also, when the switch is in the off state, current is coupled through the capacitor C_{DC} to enter the load resistance R_L. The maximum efficiency possible with a simple circuit of this type is shown [1] to be 81% when the conduction angle (indicating the time for which the switch is on) is 90° ($\pi/2$). Covering conduction angles from 0 to 180° (π), the efficiency as well as three relative power levels are shown in Figure 10.9.

In Figure 10.9 P_{lin} is the reference (linear) power level. The relative power levels are:

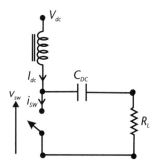

Figure 10.8 Basic concept of a switching amplifier [1]. (© Artech House, 2006.)

10.8 Harmonically Matched Power Amplifiers

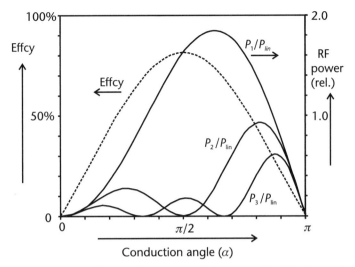

Figure 10.9 Power levels and efficiency for a basic switching amplifier [1]. (© Artech House, 2006.)

- P_1/P_{lin} is the relative power at the fundamental frequency f_0;
- P_2/P_{lin} is the relative power at the second-harmonic frequency f_2;
- P_3/P_{lin} is the relative power at the third-harmonic frequency f_3.

Observations from Figure 10.9 include:

- Efficiency peaks at 81% when the conduction angle is 90° ($\pi/2$);
- P_1/P_{lin} peaks at approximately 1.9 (around 120°);
- P_2/P_{lin} peaks at approximately 0.9 (around 150°);
- P_3/P_{lin} peaks at approximately 0.3 (around 165°).

It is evident that the second and third-harmonic levels are significant with this configuration. Cripps [1] derived the following expression for the relative level of the fundamental power:

$$P_1 / P_{lin} = \frac{8 \cdot \sin^2(\alpha)}{\pi(\pi - \alpha)} \quad (10.9)$$

where α is the conduction angle.

10.8.1.2 Operation of a Class E Amplifier

A schematic diagram showing the functioning of a Class E amplifier is shown in Figure 10.10.

In circuit of Figure 10.10, the symbol ω equals $2\pi f$ where f is the general frequency variable.

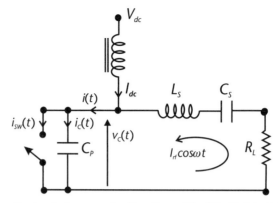

Figure 10.10 Schematic circuit diagram of a Class E amplifier [1]. (© Artech House, 2006.)

Much of this circuit is similar to that of Figure 10.8, except that an input bypass capacitor C_p is included in addition to a series tuned circuit L_s C_s inserted immediately before the load R_L. This tuned circuit implements the requirement mentioned early in this section, that is, a high-Q resonant circuit to provide a reactive load at the desired center or fundamental frequency f_0 but also implementing an open circuit (in this example) at one of the harmonics $2f_0$ and $3f_0$.

Switching currents involved with this configuration exhibit very rapid (theoretically instantaneous) transitions at peak switching times.

10.8.1.3 Example of a Specific Ultrahigh Frequency Class E Amplifier

A specific example of an ultrahigh frequency Class E Amplifier circuit is shown in Figure 10.11.

In the circuit of Figure 10.11 the following parameters apply:

- Conduction angle 125°;
- Peak current 1A;
- DC supply voltage V_{dc} = 4.8V;
- Reactance of the series inductor = 8Ω (at 850 MHz).

Simulation of this circuit reveals that the drain-source voltage V_{ds} peaks at approximately 13V and the drain current I_{ds} peaks at approximately 9 mA (at

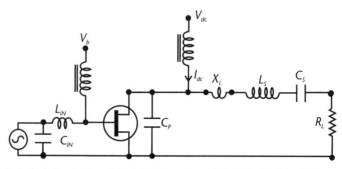

Figure 10.11 Detailed circuit of an 850-MHz Class E amplifier [1]. (© Artech House, 2006.)

different times in the cycles). The overall efficiency is 87% and the output power at the fundamental is 30.2 dBm.

10.8.1.4 Efficiency Considerations and Waveform Engineering

Designing a switched-mode microwave power amplifier with in-phase voltage and current waveforms leads to a theoretical efficiency of 123%. However, by introducing a 45° differential phase offset (Figure 10.12), the efficiency is then reduced to a more realistic 87%.

The above considerations lead to a need for waveform engineering [1] and an example of this approach is shown in Figure 10.13.

In Figure 10.13 it can be seen that the voltages and currents peak at different phases (180° difference) and the maximum voltage is approximately $2.9V_{dc}$.

Cripps's book [1] is recommended for further details on this as well as many other relevant topics.

10.8.2 Class F Power Amplifiers

10.8.2.1 Some Fundamental Aspects

In the ideal Class F configuration, all odd harmonics are terminated in open circuits, while all even harmonics are terminated in short circuits

Ideally these conditions should apply to both input and output circuitry but for most practical Class F circuits only the output (load) side is considered and a typical circuit is shown in Figure 10.14.

The circuit of Figure 10.14 and the design equations that follow appear to have originated with Chang et al. [5]. L_1 and C_1 resonate at f_0 while L_3 and C_3 resonate at $3f_0$.

In Figure 10.14 the important components, from the RF resonating aspect, are C_1, L_1, L_3, and C_3, taken in that order (C_2 is a DC blocking capacitor). Formulas developed and presented by Chang et al. [5] provide for the calculation of the two capacitors and the two inductors.

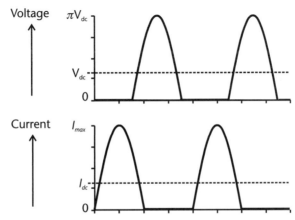

Figure 10.12 Voltage and current waveforms for a switched-mode RFPA with phase offset [1]. (© Artech House, 2006.)

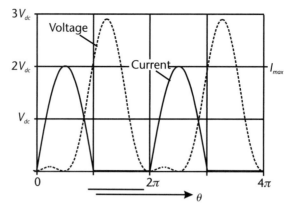

Figure 10.13 Fully realizable family of voltage and current waveforms, using second-harmonic enhancement of the voltage [1]. (© Artech House, 2006.)

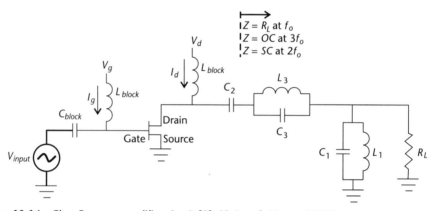

Figure 10.14 Class F power amplifier circuit [3]. (© Artech House, 2016.)

$$C_1 = \frac{E}{2\pi f_0 R_L (1-E^2)} \text{ (in farads)} \qquad (10.10)$$

in which: $E = 1 - 0.5 \frac{\Delta f}{f_0}$, Δf is the bandwidth (in hertz), f_0 is the design (center) frequency (also in hertz) and R_L is the load resistance in Ω the real part of Z_L, see (10.8).

$$L_1 = \frac{1}{4\pi^2 f_0^2 C_1} \text{ (in henries)} \qquad (10.11)$$

in which L_1 is the inductance (H), f_0 is the design (center) frequency (in hertz), and C_1 is the capacitance (F) already calculated using (10.10).

$$L_3 = 1.9753 \frac{L_1 R_L^2}{9R_L^2 + 16\pi^2 f_0^2 L_1^2} \text{ (in henries)} \qquad (10.12)$$

10.8 Harmonically Matched Power Amplifiers

in which all the parameters have already been defined.
Finally:

$$C_3 = \frac{1}{36\pi^2 L_3 f_0^2} \text{ (in farads)} \quad (10.13)$$

In this example, only the calculation of C_1 will be undertaken. It will also be assumed that $R_L = 16\Omega$. Calculate the value of capacitor for a Class F amplifier having a bandwidth of 500 MHz and a center frequency of 14 GHz.

First, calculating the E factor: $E = 1 - 0.5 \ 500/14{,}000 = 0.9821$, and then substituting into (10.10):

$$C_1 = \frac{0.9821}{2\pi 14 \times 10^9 \times (1 - 0.9821^2)} = 0.31468 \times 10^{-9} \text{ farad} = 314.7 \text{ pF}$$

In practice, the theoretical 100% maximum efficiency cannot ever be achieved. Aspects preventing this possibility include:

- The losses associated with all the passive elements determine that perfect open-circuits and short-circuits are not actually feasible.
- Properly terminating (harmonically matching) for all harmonics would demand an infinite number of resonant circuits.

Some Class F amplifiers include harmonically matching circuits on the input as well as the output, for example, terminating for $2f_0$ on the input in addition to terminating for $4f_0$ on the output.

Occasionally, a distributed approach (usually microstrip) to Class F circuit realization may be preferred to the lumped component approach as described above.

Specific designs of Class F amplifiers can handle signal bandwidths up to about 1 GHz.

10.8.2.2 Practical Example of a Class F Amplifier

Zhao et al. [6] published details concerning a broadband, high-efficiency RFPA based on the Class F mode of operation. This research team show how the basic theory of Class F operation leads to impedance-defining expressions and also to a curve (their Figure 3) indicating how the efficiency degrades on either side of the center frequency. Based on an equivalent parasitic model of the GaN HEMT used in this circuit, the load impedance (including parasitic parameters) is determined and the output matching network is designed. Radial stubs (see Chapter 3) are extensively used in the output matching network design, as shown in Figure 10.15.

In Figure 10.15 the following observations apply:

1. The cascade of four radial stubs, spaced each side of the main microstrip line, form the main matching circuit network.

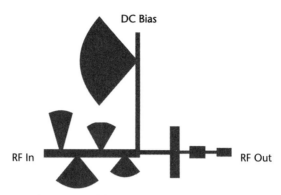

Figure 10.15 Output matching network for a Class F power amplifier [6]. (© Horizon House, 2016.)

2. The large radial stub located on the DC bias input line acts as an effective RF choke, minimizing RF energy return to the DC supply.
3. The low-high-low cascade of microstrip elements, ending with the RF output, is a lowpass filter that minimizes the transfer to the output of any unwanted signal components.

This amplifier delivers the following performance characteristics:

- Bandwidth 1.95 to 2.65 GHz (30.4% fractional bandwidth);
- RF output power 39.7 to 41.7 dBm (i.e., nominally 10W);
- PAE 60% to 79.9%.

Digital predistortion (DPD) is also applied to this amplifier. The results are shown later in this chapter, where the subjects of linearity and linearization techniques are discussed.

10.9 The Doherty Power Amplifier Configuration

In electronics generally there are precious few developments that date back in principle to pre-World War II days and yet remain highly relevant today. However, Doherty's novel power amplifier design is certainly one key example [7]. In 1936, it was 11 years before the transistor would be invented, so Doherty's realization at that time implemented electronic vacuum tubes (or valves).

Simply choosing any one of the amplifier classes described in Section 10.6 always involves a compromise, one that is often unacceptable for applications in communications systems. This compromise is between reasonably high efficiency combined with acceptably good linearity.

Doherty's basic concept was to combine two amplifiers, each having contrasting performance specifications. In this scheme, two classes of RFPAs are implemented so as to take advantage of their respective merits. An example of an overall Doherty amplifier configuration is shown in Figure 10.16.

The operation of the circuit (Figure 10.16) is as follows:

10.9 The Doherty Power Amplifier Configuration

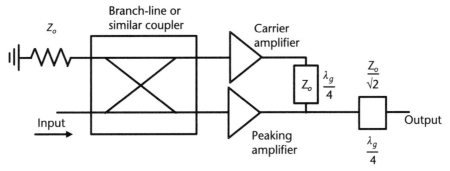

Figure 10.16 The basic Doherty amplifier configuration (based on Figure 4.22 of Kingsley and Guerci [3], with modification). (© Artech House, 2016.)

- The input signal is fed into a hybrid circuit where it is separated into two output components in quadrature (i.e., 90° phase separated) with each other.
- Each of these output components is then fed into an amplifier operating in a specific class (Class AB for the carrier amplifier and Class C for the peaking amplifier).
- The output from the carrier amplifier feeds into a quarter-wave transforming section of transmission line, when it then joins the output of the peaking amplifier at what is sometimes termed the common node.
- Finally the signal from this common node is matched by means of a final quarter-wave transformer (characteristic impedance $Z_0/\sqrt{2}$) which matches through to the output.

The carrier and peaking amplifiers are occasionally named the main and auxiliary amplifiers respectively. Hybrid circuits are described in Chapter 4 and an important advantage of this choice of circuit is the cancellation of unwanted reflected signals at its input.

At back-off only, the Class AB carrier amplifier is operating, while the Class C peaking amplifier is turned off. This Class C peaking amplifier begins to function when the signal reaches the power level at which the carrier amplifier output starts to decrease. The power transfer characteristics exhibited in Figure 10.17 illustrate the contributions of each subamplifier and the overall effect.

The quarter-wave matching lines can be replaced with more sophisticated matching circuitry so the amplifier bandwidth is thereby broadened.

The final output characteristic follows that of the carrier amplifier, until just before saturation, when the peaking amplifier begins to take over so the remaining portion of the characteristic follows the broken line Doherty curve.

The overall operation combines reasonably high efficiency as well as an acceptable degree of linearity.

References [8, 9] provide information on interesting Doherty amplifier advances.

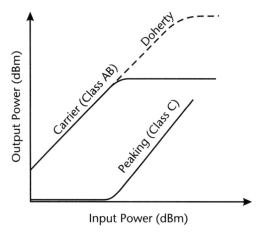

Figure 10.17 Power transfer characteristics for the Doherty amplifier [3]. (© Artech House, 2016.)

10.10 The Envelope-Tracking Amplifier

The envelope-tracking amplifier approach is radically different from that of Doherty due to digital control and the implementation of a dynamic drain supply for the internal power amplifier.

In common with the Doherty carrier amplifier, however, this internal power amplifier can be operated in either Class A or Class AB mode. The overall architecture is shown in Figure 10.18.

Figure 10.18 clearly shows the main overall architecture of this type of amplifier. The envelope amplifier is fed conventionally by the fixed-voltage DC drain supply (V_{DD}), which may be provided by a battery source, solar, or rectified from AC mains. However, this envelope amplifier is also controlled by a signal magnitude x delivered by the digital signal processor (DSP) (see Chapter 5). This causes the output voltage from the envelope amplifier to act as a dynamic drain supply to the FET-based main amplifier (PA). The other two outputs from the DSP are the imaginary and real parts of signal x and they both control a modulator and upconverter (Chapters 13 and 14) and the output from this block is effectively the input to the PA.

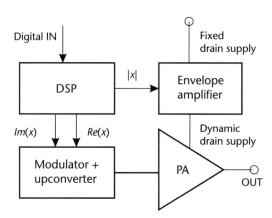

Figure 10.18 Architecture of an envelope-tracking amplifier [2]. (© Artech House, 2016.)

The overall result is to (as for Doherty) maintain both high efficiency and good linearity. Camarchia et al. [2] described this in more detail.

The approach of feeding an amplifier with a dynamic power supply has proved to be popular in several recent developments.

10.11 High Power Push-Pull Amplifiers

Push-pull amplifiers using the balanced amplifier approach are also popular in various RF applications. These types of amplifiers are described in detail in Section 8.6.2 of Chapter 8 and therefore only some brief comments regarding hybrid, high-power balanced amplifier realizations are made here.

The implementation of Lange couplers into balanced amplifier designs leads to at least one significant issue, namely a small but important degree of unbalance between the branches. This can mean that one of the two transistors will saturate sooner than the second device.

For high-power push-pull balanced microwave amplifiers, anti-phase power splitters (and combiners) or baluns are required, and the design of this type of circuit configuration is challenging. However, applying EDA (NI AWR's AXIEM EM simulation package, for example), high-power GaN HEMT based 200-W amplifiers can be designed to cover a bandwidth of at least 50% at microwave frequencies.

10.12 Other Practical RFPA Circuits

Although they are of key importance, RFPAs focused mainly on power output and efficiency are by no means the only RFPA considerations for communications systems. For example, relatively low-power, high-frequency, and moderate-efficiency RF amplifiers are required for the increasingly important short-range radio systems.

As indicated earlier in this chapter, millimeter-wave power amplifiers are of increasing significance (see Figures 10.3 and 10.4).

Also, the distributed amplifier, described in Chapter 8, is a type of low-to-moderate output power circuit that can cover extremely wide bandwidths.

The MMIC PA comprises:

- A total 12 FETs;
- Three spiral lumped inductances providing much of the matching at the input;
- Two spiral lumped inductances providing much of the matching at the output;
- Several arrays of microstrip elements.

Chapter 3 provides information on all the above types of components.

For this circuit, Cripps [1] estimated of the linear output power show approximately +31 dBm from 9 to 11 GHz.

10.12.1 Ka-Band PA MMIC Examples

The first example of a Ka-band PA cited here is one due to Plextek RFI, fully described in a YouTube video by Stuart Glynn of that company [10]. This is a dual-band millimeter-wave PA designed for the forthcoming 5G systems. The lower band is centered on 26 GHz while the higher band is centered on 32 GHz.

Early in this project, it was found that single, broadband PAs designed to include both bands always had specification disadvantages compared with the dual-band approach. Therefore, the option of implementing two, electronically switched PAs, one covering the low band (26 GHz) and the other covering the high band (32 GHz) is decided upon. This electronic switching is the key to the successful implementation of this arrangement. A commercially available 0.15-μm GaAs pHEMT process is adopted.

Capacitances and microstrip transmission lines are the only passive components implemented, with inductors having excessive parasitics at these millimeter-wave frequencies.

The use of a GaAs pHEMT as a switch is shown in its simplest realization in Figure 10.19.

The equivalent resistance of the switch is approximately 3Ω and the equivalent capacitance is 120–130 fF. The gate-source voltage V_{gs} is set at 0V for the on state and about –2.0V for the off state. These switches are positioned to switch, in and out, various sections of transmission lines and shunt capacitors so as to automatically reconfigure the circuit for operation in each band. Plextek RFI designed a novel complementary (inverting) control circuit delivering either –5V or –0.6V to operate these switches.

The overall amplifier is a three-stage design and the final output stage comprises four transistors all paralleled to obtain maximum possible output power. Average output power is >30 dBm across both bands and the gain is 19.5 dB. The efficiency is >30% across the lower band and slightly lower than this in the high band. Shinjo et al. [11] describes a three-stage GaAs-based 29 GHz EFFA.

10.12.1.1 Alternative GaN HEMT Realization of a Ka-Band Amplifier

Yamaguchi et al. [12] reported a 0.15-μm GaN HEMT MMIC operating in Ka-band. This amplifier achieved an output power of 20W, with accompanying 19%

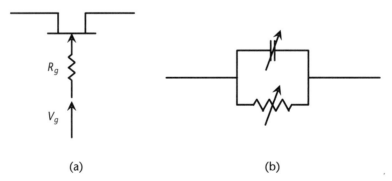

Figure 10.19 Basic concepts of a GaAs pHEMT in switch mode: (a) device with gate resistor Rg and control voltage Vg, and (b) equivalent circuit.

efficiency under CW operation at 26.5 GHz. The work is summarized in the article by Ma et al. [13].

10.13 The Distortion Issue and Linearization Techniques

10.13.1 Linearity and Intermodulation Distortion

To understand the process and the terminology associated with this subject, it is necessary to refer back to Figures 2.15, 1.20, and 10.4. All these diagrams amply demonstrate the nonlinearities inherent in all types of RFPAs. Figure 1.20 shows how these nonlinearities originate in the transistor device itself, while Figure 10.4 indicates the straight line that can be drawn, asymptotic to the P_{out} curve as P_{out} approaches zero.

In this section, the topic of nonlinearity is considered in more detail through to determining the IMD mixing products and an important equation enabling the power level associated with the third-order IMD product to be determined.

The analysis begins by observing that any amplifiers' voltage transfer function can be expressed as the following Taylor series:

$$v_{out} = a_1 v_{in} + a_2 v_{in}^2 + a_3 v_{in}^3 + a_4 v_{in}^4 + \ldots \tag{10.14}$$

It is the situation where there are two input frequencies (or tones) that is of particular interest because this is how the distortion is seen. As usual it is mathematically most convenient if these two inputs, operating at frequencies f_1 and f_2, are chosen to be cosine waves, as follows:

$$v_{in}(t) = \cos(2\pi f_1 t) + \cos(2\pi f_2 t) \tag{10.15}$$

Substituting this expression into (10.14) and curtailing the resulting expression at the v_{in}^2 term, gives

$$v_{out} \approx a_1 \{\cos(2\pi f_1 t) + \cos(2\pi f_2 t)\} + a_2 \{\cos(2\pi f_1 t) + \cos(2\pi f_2 t)\}^2 \tag{10.16}$$

The distortion effect lies within the term having coefficient a_2, which can be expanded out and then the individual frequency components can be identified after using trigonometric identities, yielding the term

$$\begin{aligned} a_2 \{&1 + 0.5\cos(4\pi f_1 t) + 0.5\cos(2\pi(f_1 - f_2)t) \\ &+ 0.5\cos(2\pi(f_1 + f_2)t) \\ &0.5\cos(4\pi f_2 t)\} \end{aligned} \tag{10.17}$$

These four frequency terms include double the frequencies of the original two tones, that is, $2f_1$ and $2f_2$ but also and very importantly two further signal components at frequencies $(f_1 - f_2)$ and $(f_1 + f_2)$. These are termed mixing products or, occasionally, spurious signals.

By analyzing the third term in (10.14) (i.e., with coefficient a_3) further mixing products are revealed at frequencies: $3f_1$, $3f_2$, $2f_1 + f_2$, $f_1 + 2f_2$, $2f_1 - f_2$ and $2f_2 - f_1$. These results can be conveniently displayed on a spectrum diagram but instead a numerical example of K-band amplifier mixing products is given (Table 10.1).

Clearly, from Table 10.1, there are two mixing product frequencies, both of which are quite close to the input tones (i.e., 26.999 GHz and 27.002 GHz). In practice, these would be extremely hard to filter out, demanding filters with very sharp skirts (see Chapter 6).

For an amplifier (or other circuit component exhibiting IMD), the various IMD components are characterized by distinctly different slopes of their graphical representations: second-order 2 dB for every 1-dB increase in input power, third-order 3 dB for every 1-dB increase in input power.

Conventionally, an output/input power transfer characteristic will be associated with the following linearity terms:

- IIP2 = second-order input intercept point;
- IIP3 = third-order input intercept point;
- OIP2 = second-order output intercept point;
- OIP3 = third-order output intercept point.

All these terms as well as the characteristic slopes are indicated in Figure 10.20.

The most frequently used IMD specification term relates to IM3 and the relative level of this can be calculated using the following identities:

$$IM3 \ (dBc) = \frac{P_{2f_2 - f_1}}{P_{f_2}} = \frac{P_{2f_1 - f_2}}{P_{f_1}} \quad (10.18)$$

For general guidance, OIP3 tends to be around 10 dB above the value of P_{sat}, which is a useful general rule.

In practice, specifications provided by suppliers can be of suspect value and the best approach is to measure the actual IM values.

An important additional note is that IMD is not only an issue with amplifiers; it is also present in situations as basic as junctions between media and various

Table 10.1 Mixing Products Generated from Two Input Tones at 29 GHz and 29.001 GHz (All Data in Gigahertz)

First	Second	Third
$f_1 = 29$	$2f_1 = 56$	$3f_1 = 87$
$f_2 = 29.001$	$2f_2 = 56.002$	$3f_2 = 87.003$
	$f_1 - f_2 = 0.001$	$2f_1 + f_2 = 141.001$
	$f_1 + f_2 = 58.001$	$f_1 + 2f_2 = 87.002$
		$2f_1 - f_2 = 26.999$
		$2f_2 - f_1 = 27.002$

10.13 The Distortion Issue and Linearization Techniques

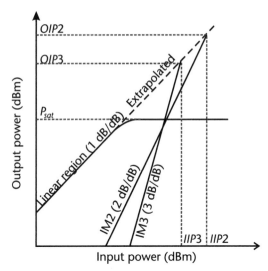

Figure 10.20 Power transfer characteristics for an RFPA indicating the various terms resulting from the analysis of nonlinearity [3]. (© Artech House, 2016.)

other practical passive circuits. Knowledge concerning IMD is essential in many circumstances and must be reduced to levels as low as possible.

10.13.2 Linearization Techniques

It is possible to introduce special circuits that will greatly improve the overall linearity of an RFPA module. The basic approach, termed predistortion, is indicated in Figure 10.21.

The block diagram of Figure 10.21 illustrates, in signal-flow form, how predistortion works. The signal first enters the basic amplifier (which will be almost any of the types described in this chapter), but this amplifier has an inherent distorting power transfer characteristic. The output of this basic amplifier is fed to the predistorter, which is designed to exhibit as exactly as possible the opposite power transfer characteristic to that of the basic amplifier, thereby canceling the distortion and leading to the final result shown.

Practical realizations comprise either analog or digital predistorters. An analog predistorter typically implements a loop including a diode that detects when the output power level is tending towards nonlinearity and then corrects by feeding

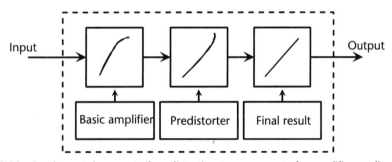

Figure 10.21 Fundamental concept of predistortion to compensate for amplifier nonlinearity.

through either less power or (usually) more power. An arrangement implementing baseband digital predistortion (DPD) is shown in Figure 10.22.

Preknowledge of the basic PA input-output power transfer characteristic is always required. In the digital predistortion scheme shown in Figure 10.22 the output voltage from the DPD, normally operating at IF, is fed into two DACs that, in turn, drive the upconverter. The output of this upconverter operates at the required microwave or millimeter-wave frequency and it is followed by the final PA. More details have been provided by Camarchia et al. [2].

The effect of applying DPD to a real, practical amplifier is best illustrated by means of its output power spectra and an example is given in Figure 10.23.

Clearly, from Figure 10.23, the effect of applying DPD is to reduce the output power over close-in bands each side of the 2.14-GHz carrier while maintaining overall symmetry. In particular, over the frequency range 2.133 GHz to 2.138 GHz, the application of DPD causes the output power to be depressed by 15 dB. At frequencies further away from the carrier, there is still an improvement of around 2 or 3 dB.

10.14 Some Final Overall Comments Regarding RFPAs

This chapter covers a substantial range of topics relating to RFPAs. All of these topics are important considerations for informed device (transistor) selection followed

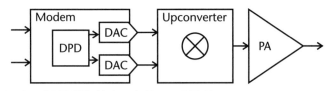

Figure 10.22 Baseband DPD [2]. (© Artech House, 2016.)

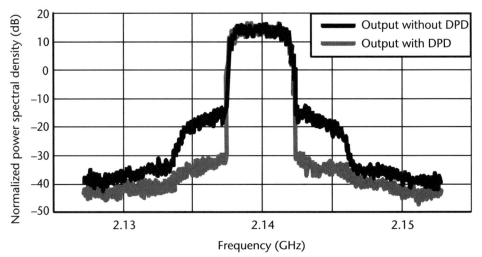

Figure 10.23 Output power spectra for a Class F amplifier with and without DPD. (Courtesy of Zhao et al. [6].)

by choice of power amplifier configuration and design. However, there are two strategic aspects that are both (taken together) potentially exciting regarding the future for RFPAs. These are:

- The fact that the gain of antennas increases with operating frequency (see Chapter 7).
- The fact that available RF output power from RFPAs decreases with operating frequency (see Figures 10.1, 10.2, and 10.3).

There are many technological developments and plans concerning 5G scenarios. Several of the references cited here relate to these developments. For the full realization of 5G, millimeter-wave technology is vital but will, mainly at least, not require moderate-to-high power RFPAs. This is because relatively short-range links are envisioned through what has become known as network densification. As a direct result, the two strategic aspects above will become increasingly significant.

References

[1] Cripps, S., *RF Power Amplifiers for Wireless Communications*, Norwood, MA: Artech House, 2006.

[2] Camarchia, V., R. Quaglia, and M. Pirola, *Electronics for Microwave Backhaul*, Norwood, MA: Artech House, 2016.

[3] Kingsley, N., and J. R. Guerci, *Radar RF Circuit Design*, Norwood, MA: Artech House, 2016.

[4] Grebennikov, A., and N. O. Sokal, *Switchmode RF Power Amplifiers*, New York: Elsevier-Newnes, 2007.

[5] Chang, K., I. Bahl, and V. Nair, *RF and Microwave Circuit and Component Design for Wireless Systems*, New York: John Wiley & Sons, 2002.

[6] Zhao, Z., et al., "Broadband High Efficiency Power Amplifier Design Using Continuous Class F Mode," *Microwave Journal*, May 2017, pp. 132–146.

[7] Doherty, W., "A New High Efficiency Power Amplifier for Modulated Waves," *Proc. IRE*, Vol. 24, No. 9, September 1936, pp. 1163–1182.

[8] Wong, J., N. Watanabe, and A. Grebennikov, "Efficient GaN Doherty Amplifier Peaks at 1 kW from 2.11 to 2.17 GHz," *Microwaves & RF*, May 2017, pp. 64–69 and 149.

[9] Cho, Y., et al., "A Handy Dandy Doherty PA," *IEEE Microwave Magazine*, September/October 2017, pp. 110–124.

[10] https://youtube/HjYCLoIEdsA.

[11] Shinjo, S., et al., "Integrating the Front End," *IEEE Microwave Magazine*, July/August 2017, pp. 31–40.

[12] Yamaguchi, Y., et al, "A CW 20W Ka-Band High Power MMIC Amplifier with a Gate," *Proc. IEEE Compound Semiconductor Integrated Circuit Symposium*, 2017, pp. 1–4.

[13] Ma, R., et al., "A GaN PA for 4G LTE-Advanced and 5G," *IEEE Microwave Magazine*, November/December 2017, pp. 77–85.

CHAPTER 11

RF-Oriented ADCs and DACs

11.1 Introduction

Several types of ADC architectures are available and those that are particularly applicable to RF-oriented ADCs are described in this chapter. All types of ADCs represent fairly complex components and most are therefore implemented as MMICs or RFICs. Indeed, both ADCs and DACs are increasingly implemented as subsystem circuits within a more comprehensive chip, with many ADCs and DACs on the same chip.

Specialized requirements such as ultrahigh-precision converters (typically for instrumentation) often require realization in the form of hybrid circuits using discrete components.

11.2 ADCs

11.2.1 Quantization and Sampling

The original analog signal must be processed in some manner such that the required digital output signal is produced. The processing requires sampling the voltage levels of the analog signal at regular intervals. At first, it may be thought that all levels of the input analog signal, as this progresses with time, should be sampled, but this would demand sampling at zero-spaced intervals, which is totally impracticable to realize. Even sampling at time intervals very much smaller than the analog waveform period is excessive and cannot be realized with practical arrangements.

The actual conversion process used is termed quantization of the input signal.

All types of ADC perform the quantization process on a periodic basis: sampling the input on a regular basis. Sampling results in the continuous nature of the analog input becoming what is termed a discrete-time digital output. Also, this essential quantization produces specific kinds of errors which are however quantifiable and will be discussed in some detail here. The quantizing and sampling processes are indicated in Figure 11.1.

To directly quote Mathuranathan [1], "Nyquist-Shannon Sampling Theorem is the fundamental base over which all the digital processing techniques are built."

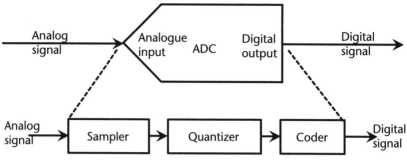

Figure 11.1 The ADC processing sequence. (After: [1].)

An understanding of the sampling theorem is therefore vital in order to advance any further regarding an understanding of ADCs. The sampling process first records the magnitude of the input analog signal at regular intervals and the rate at which this proceeds is known as the sampling rate, the time interval between samples being known as T_s. Correspondingly, the sampling frequency, denoted f_s, is therefore

$$f_s = 1/T_s \qquad (11.1)$$

The important question (which the Nyquist-Shannon sampling theorem helps answer) is: How does the ADC designer choose the sampling rate or frequency such that the original analog signal is faithfully reproduced in the digital domain? To answer this question, it is necessary to check the fundamental version of the Nyquist-Shannon sampling theorem, which applies to baseband analog sampling. This version of Nyquist's theorem states: "In order for a faithful reproduction and reconstruction of an analog signal that is confined to a maximum frequency f_m the signal should be sampled at a sampling frequency f_s that is greater than or equal to twice the maximum frequency of the signal."

Mathematically expressed as (11.2):

$$f_s \geq 2f_m \qquad (11.2)$$

This is an extremely important result that must be adhered to and that fundamentally impacts all digital signal processing.

Example 11.1

Consider a baseband analog signal that extends from (effectively) DC to 12 kHz (f_m). In order to faithfully convert this signal to the digital domain, what is the minimum frequency at which the sampler should operate?

Applying (11.2), the result is clearly $f_{s,min} = 24$ kHz.

This means that the input analog signal could be sampled at a frequency of 26 kHz, for example, just 2 kHz above the 24-kHz minimum.

There is also bandpass sampling in which the analog input frequency ranges across a defined bandwidth. Binghampton [2] provided an excellent resource on this aspect of ADCs.

11.2.2 Sampling in Practical ADCs

In practice, the sampling is applied by means of a sampling clock and an example of how this can be achieved using a symbol clock recovery feedback loop is shown in Figure 11.2.

In Figure 11.2, the basic ADC circuit is clocked with an input wave operating at a frequency exceeding twice the maximum signal input frequency. The clock (or sampling) waveform is obtained by selecting the higher-frequency channel emerging from the ADC output, feeding a fraction of this through a lowpass filter and using the output voltage of this filter to drive the VCO sampling clock. Filter design is covered in Chapter 6 and VCOs are described within Chapter 12.

However, a drawback with this configuration is the substantial length of the feedback loop and this factor can degrade the performance. As a result, alternative methods have been sought and an example is shown in Figure 11.3.

In Figure 11.3, the basic ADC, sampling clock, and channel selection filter are essentially identical to those shown in Figure 11.2. The important difference is that the DSP is now more sophisticated. Following the channel selection filter, there is an interpolator circuit that is driven by a numerically controlled oscillator (NCO), which is effectively the digital equivalent of an analog voltage controlled oscillator (VCO).

11.2.3 Effective Number of Bits

Before proceeding further, it is essential to define and explain the important concept of the effective (or equivalent) number of bits (ENOB) and how this parameter impacts the characteristics and design of ADCs.

The major parameters defining any ADC are its bandwidth (up to several gigahertz achievable commercially), its dynamic range, and its signal-to-noise ratio (SNR) in decibels. The bandwidth is mainly determined by the sampling rate.

The dynamic range is affected by the ADC's accuracy, linearity, and resolution as well as aliasing and jitter (considered later here). Also, the dynamic range is often measured in terms of the ENOB. The signal-to-noise ratio is also a function of the ENOB as well as other critical parameters. The ENOB is the number of bits that each sampled measure produces that originate from the true analog signal (i.e., are not due to noise).

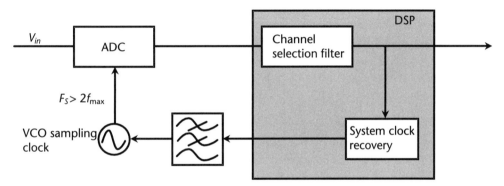

Figure 11.2 VCO-based sampling clock driven by the recovered (and filtered) symbol clock. (From: [3]. © Artech House, 2016.)

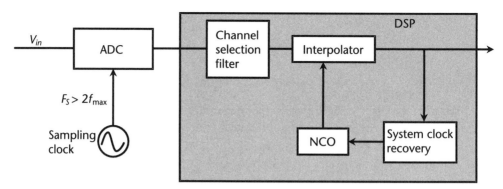

Figure 11.3 ADC implementing a free-running VCO-based sampling clock with numerical adjustment. (From: [3]. © Artech House, 2016.)

However, an issue known as quantization error gives rise to quantization noise, which degrades the signal-to-noise performance of any ADC. (Quantization error is discussed later here.)

Another parameter associated with an ADC is the number of bits N_b, measured in each sample. The minimum value of N_b determines the smallest variation in analog signal voltage that can be resolved. The quantization process inherently results in a source of noise and taking this noise alone has been shown to yield a signal-to-(quantization)-noise ratio given by [4]:

$$\left(S_{av}/N_q\right) = 6.02 N_b + 1.76 - \left(S_{peak}/S_{av}\right) + 3 \ \text{dB} \tag{11.3}$$

where S_{av} and S_{peak} are, respectively, the average and peak signal power levels, N_q is the average quantization noise power, and N_b is defined above. The numerical quantities: 6.02, 1.76, and 3 are explained in Section 11.2.5.

$\left(S_{peak}/S_{av}\right)$ often simply termed PAPR, is 3 dB for a purely sinusoidal waveform but will be as high as 18 dB for a 1024QAM modulated signal.

Example 11.2

An ADC operates with an average signal-to-noise ratio of 45 dB and a PAPR of 18 dB. What is the ENOB?

Rearrange (11.3) for N_b and substitute these data:

$$N_b = (45 - 1.76 + 18 - 3)/6.02$$

which calculates to $N_b = 9.67$.

In practice, this would be rounded to 10. ENOB values between around 5 and 12 are commonly encountered in ADC designs.

Closely related noise effects in the analog sections of radio receivers are considered in Chapter 9, which focuses on LNAs and associated receiver design parameters.

11.2 ADCs

Incidentally, noise may be expressed either as noise power directly (W, or equivalently dBm) or as volts per $\sqrt{Hz}$. Why volts per $\sqrt{Hz}$? Fundamentally, noise power is given by kTB where k is Boltzmann's constant (1.38×10^{-23} J.K^{-1}), T is absolute temperature (K), and B is the bandwidth in hertz. Where this noise emanates from a resistance R and bearing in mind the voltage divider effect, then the noise power is also given by $v_n^2/4R = kTB$ and therefore v_n is proportional to $\sqrt{B}$ (i.e., the units are volts per $\sqrt{Hz}$).

11.2.4 Quantization Error and Quantization Noise

As described above, the major processes with an ADC are sampling and quantization. It is vital to recollect that these processes are essentially discrete in nature (i.e., certainly not continuous).

If the (binary) output code is plotted against the input signal voltage, then the relationship shown in Figure 11.4 results (due to quantization).

An expression for the overall signal-to-(quantization)-noise ratio can be found by considering the following expressions relating to quantization error (Q), input (voltage V_m) signal range, variance P_s, and peak power P_{peak}. First, this leads to the quantization error variance P_q [3]:

$$P_q = \frac{P_{peak}}{3.2^{2N_b}} \qquad (11.4)$$

which, in turn, develops into the signal-to-(quantization)-noise ratio:

$$\left(S_{av}/N_q\right) = \frac{P_s}{P_{peak}} 3 2^{2N_b} \qquad (11.5)$$

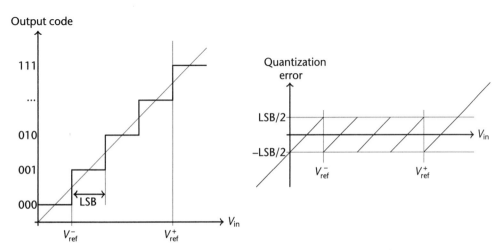

Figure 11.4 Ideal ADC output code versus input voltage staircase waveform (left) and quantization error (right). (© Artech House, 2016.)

which finally leads to the following expression for signal-to-(quantization)-noise ratio, applying to uniformly distributed quantization noise:

$$\left(S_{av}/N_q\right)_{uniform} \approx 6.02 N_b \text{ dB} \qquad (11.6)$$

which is the first term in (11.3), in which the peak-to-average power ratio and the additional noise of 3 dB for a sine wave input are both included. This quantity is sometimes termed the signal-to-quantization-noise ratio (SQNR).

11.2.5 Quantization Static Error and Sampling Distortion

As a signal progresses through an ADC, it is degraded by various sources of errors at least one of which is caused by the quantization process itself. Principally, as a result of very small differences in structure manufacturing irregularities across the semiconductor die, the quantization process will not be perfectly linear. This can be seen by studying the output binary code versus stepped voltage characteristic which is illustrated in Figure 11.5.

The ideal linear response is indicated by the faint response, whereas the real linear response is indicated by the bold-line response. Clearly, in the real response the steps are nonuniform and this nonuniformity accounts for the quantization errors.

In Figure 11.5, the nonuniformity is exaggerated to show the effect clearly.

All electronic operations require finite time and quantization is no exception. It is therefore necessary for the sampling period to be sustained so that sampling can be properly completed within the time for quantization and most often in ADCs a track-and-hold operation supplies this requirement. An example of a relatively simple track-and-hold circuit is shown in Figure 11.6.

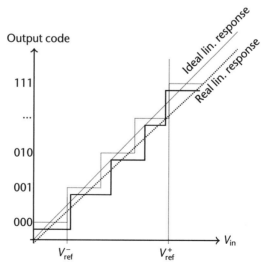

Figure 11.5 ADC output code versus input voltage staircase responses: ideal and real. (© Artech House, 2016.)

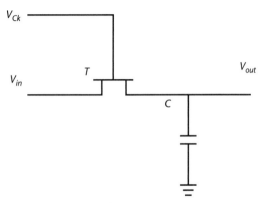

Figure 11.6 Track-and-hold internal circuit for an ADC.

In Figure 11.6, V_{Ck} is the clocking input, driving the MOS switching transistor T. V_{in} is the signal input, transferable to the output but including the important discharging capacitor C.

During the tracking portion of the process, the voltage across the capacitor tracks the input (as the current flows through the relatively small on resistance of the MOS transistor). When the clock causes the switch to open, the previously stored charge within the capacitor C decays a little but still delivers an approximately constant voltage to the output. The small on resistance of the MOS transistor arises from the device's channel and leads to a small but influential nonlinearity. Because of this high-speed (including RF) ADCs implement somewhat more complex switching arrangements. Related charge-storage effects cause some nonlinear distortion during the switch opening portion of the process.

Typical track-and-hold waveforms are shown in Figure 11.7.

The uneven steps in the upper waveform (voltage V) indicate sampling distortion.

The on resistance associated with the MOS switch also contributes thermal noise and a simple equivalent circuit representing this noise is shown in Figure 11.8.

In Figure 11.8, V_n is the equivalent noise voltage source, R_{on} is the MOS transistor on resistance, C is the charge storage capacitance (as defined above) and $V_{n,out}$ is the output noise voltage. This noise is actually independent of the resistance value but varies inversely with the capacitance, C, following the expression kT/C. In the final design of the sample-and-hold internal circuit, the value of the capacitance C is kept low enough for effective sampling (and low-power consumption) yet high enough to keep the noise lower than the quantization noise level.

11.2.6 Sampling Jitter

This phenomenon has a highly significant effect on the performance of any ADC; in particular, it can seriously degrade the ENOB.

Sampling jitter originates in the inevitable phase noise associated with the clock oscillator. All oscillators are subject to phase noise and this is considered in Chapter 12. This jitter in the sampling clock perturbs the desired sampling period, shift-

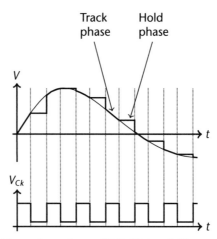

Figure 11.7 Track-and-hold waveforms for an ADC. (© Artech House, 2016.)

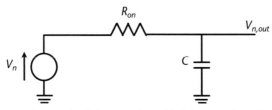

Figure 11.8 Equivalent noise circuit of the sample-and-hold internal circuit of Figure 11.6.

ing this period by a small but significant amount of time. The effect is shown in exaggerated form in the waveforms of Figure 11.9.

Clock phase noise and jitter have been discussed in some detail by Neu [5].

The maximum rate of change is directly related to the maximum (spectral) frequency associated with the input signal and this leads to the maximum signal-to-noise ratio due to jitter. The expression for this ratio is [6]:

$$(S/N)jitter = \frac{1}{4\pi^2 f^2 t_{jitter}^2} \tag{11.7}$$

(This quantity is frequently termed the signal to aperture jitter noise ratio [SJNR].)

From this equation, it can be seen there is an inverse frequency-squared behavior and, because ENOB is closely related to this (S/N)*jitter* behavior this means that ENOB is also a function of input frequency, decreasing steadily as frequency increases. Figure 11.10 gives a general indication of the typical variation for a high-frequency ADC.

In practical ADCs, ENOB will not be a purely straight-line function with increasing frequency. Instead, various highly nonlinear noise and distortion effects will result in a distinct downward trend but also exhibiting small fluctuations.

11.2 ADCs

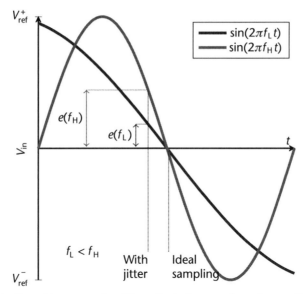

Figure 11.9 Illustration of how sampling jitter affects the sampling of a specific signal waveform. (© Artech House, 2016.)

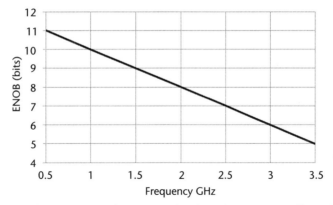

Figure 11.10 Typical ENOB versus frequency plot (based on a nominally 12-bit ADC in this instance).

11.2.7 Aliasing and Antialiasing

The effects of unwanted (usually interfering) signals can result in serious performance disturbances for ADCs and must be identified followed by ideally elimination, in practice minimization down to an acceptable level.

To see the effects of this aliasing, consider two incoming frequency components: one the desired signal of frequency f_1 (= 10 MHz) but also an unwanted component f_2 (= 20 MHz). Sampling is to take place at 30 MHz. The 10-MHz desired frequency component will generate the following spectral components at the output of the sampler:

$$10 \text{ MHz, } 20 \text{ MHz, } 40 \text{ MHz, } 50 \text{MHz, } 70 \text{ MHz, ...}$$

The unwanted frequency component ($f_2 = 20$ MHz) will cause the following spectral components to appear at the output of the sampler:

20 MHz, **10 MHz**, 50 MHz, 40 MHz, 80 MHz, ...

The important point to observe is the reemergence of a 10-MHz component, directly interfering with the original (desired) 10-MHz signal. This is termed the alias of the original signal and both are indicated in bold among the above sequences of frequency components. This alias of the original signal is indistinguishable from the original (desired) signal. The overall situation, including signal spectra, is shown in Figure 11.11.

In a similar manner, the 20 MHz generated by the original $f_1 = 10$ MHz signal is an alias of the original 20-MHz component and this alias will interfere with the original. However, ultimately only the converted 10-MHz signal is required and all frequencies exceeding this value can be filtered out.

It is still necessary to take care of the aliasing 10-MHz component generated as a result of the 20-MHz unwanted signal. The key toward overcoming this problem lies in the fact that aliasing depends on the sampling frequency and its relationship with the frequency components involved with the ADC.

If a signal is sampled at its own frequency f_s, then all the components ranging from $f_s/2$ through to f_s will be aliases of frequency components ranging from 0 to $f_s/2$ and also the reverse. This particular frequency $f_s/2$ is termed the folding frequency because the frequency components ranging from $f_s/2$ up to f_s fold back on themselves and interfere with the components ranging from 0 to $f_s/2$ and vice

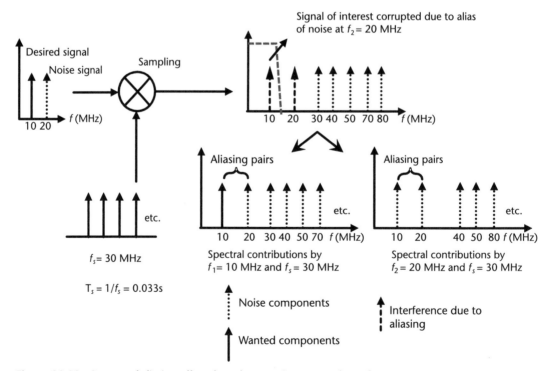

Figure 11.11 Spectra of aliasing effects based on two input megahertz frequency components.

versa. Aliasing zones exist on both sides of the spot frequencies: $0.5f_s$, $1.5f_s$, $2.5f_s$, $3.5f_s$, $4.5f_s$, ...

All the frequencies quoted in the above sequence are termed folding frequencies because all are associated with frequency reversal.

On a similar basis, aliasing zones are also present either side of the following spot frequencies:

$$f_s, 2f_s, 3f_s, 4f_s, 5f_s, ...$$

but with no frequency reversals.

The concept of aliasing zones is illustrated in Figure 11.12.

In Figure 11.12 zone 2 is a mirror image of zone 1 but with frequency reversal. Zone 2 will generate aliases in zone 3, although without frequency reversal. Next, zone 3 generates a mirror image in zone 4 with frequency reversal, and so on.

Referring to the numerical example used earlier, here the folding frequency was found to be $f_s/2$ where f_s is the sampling frequency, which was 30 MHz. Therefore, $f_s/2$ is 15 MHz, which means that all the spectral components ranging from 15 MHz to 30 MHz will be aliases of those components ranging from 0 Hz to 15 MHz.

These unwanted aliasing components will become present within the required band, making it impossible to identify the actual signal. Therefore, it is essential to introduce a filter (termed an antialiasing filter) ahead of the ADC. This antialiasing filter must be designed to remove all frequency components lying above the folding frequency (i.e., above 15 MHz in the above example). This calls for a lowpass filter (LPF) with a cutoff frequency set just below the 15-MHz frequency. The general arrangement is shown in Figure 11.13.

11.2.8 Adjacent Channel Power Ratio

Digital signal modulation, dealt with in some detail in Chapter 14, can lead to an unwanted phenomenon known as spectral regrowth. This phenomenon is caused

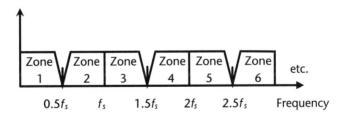

Figure 11.12 Folding frequencies and aliasing zones.

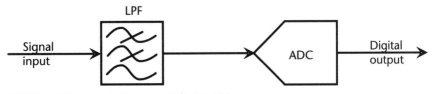

Figure 11.13 ADC preceded by an antialiasing LPF.

by distortion and the full output spectrum now takes on the appearance of an undesired out-of-band noisy image of the original signal.

The ratio between the peak power levels of the original signal pulses and those associated with the spectrally-regrown pulse is known as the adjacent channel power ratio (ACPR). This ACPR parameter can be used toward the design of antialiasing filters. For this, it is necessary that the frequencies associated with the unwanted spectrally regrown pulse be accurately measured so that the filter's frequency response will include those frequencies. The design of appropriate filters is covered in Chapter 6.

11.3 ADC Architectures

Many types of ADC architectures exist, although only a few meet the stringent specifications required for RF signal conversion. In this section, several appropriate architectures are considered.

11.3.1 The Flash ADC Architecture

This type of ADC exhibits the lowest latency of all configurations because it principally comprises a stack of parallel comparators (usually implemented with operational amplifiers [7]). The overall architecture is shown in Figure 11.14.

It is necessary to have 2^{N_b} comparators, each having increasingly high-voltage thresholds progressing upwards according to the set of series resistors with reference voltages as indicated in Figure 11.14. Relating to any particular sample the comparator with the highest output is selected using the priority encoder, the delivered output digital code. It is necessary for the decoding circuit to be designed to minimize the errors that typically occur as a result of monotonic outputs, known as bubbles, which can appear close to the required sampled value [8].

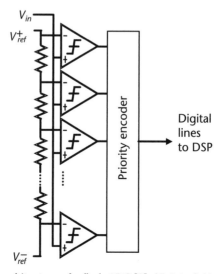

Figure 11.14 The overall architecture of a flash ADC [3]. (© Artech House, 2016.)

Because of the all-parallel configuration, the conversion from the analog signal to the digital output is essentially single-step and, for an 8-bit ADC, the process is completed within some tens of nanoseconds. This relatively high operating speed comes at the cost of various voltage drifts and digital waveform uncertainties that lead to relatively poor linearity.

11.3.2 The Folding ADC Architecture

This overall type of architecture comprises two ADCs: one dealing with the most significant bit (MSB) of the conversion process, and one handling the least significant bit (LSB) of the conversion process. The concept is aimed at improving the resolution of a flash ADC while retaining its inherently low latency and the architecture is shown in Figure 11.15.

In this circuit arrangement, the coarse ADC handles the MSB of the conversion process while the fine ADC handles the LSB of the conversion process. The folding circuit wraps the input signal so that its output bit stream corresponds to the LSB of the coarse ADC. The analog input signal is delayed before entering the coarse ADC, and this delay matches closely that introduced by the folding circuit so the sampling times are synchronized at both ADC input ports.

11.3.3 Pipelined ADC Architecture

The overall arrangement of a pipelined ADC is shown in Figure 11.16. An alternative name for this type of ADC is a subranging quantizer.

The lower diagram in Figure 11.16 represents any one of the m-bit stages present in the comprehensive pipelined ADC shown in the upper diagram. In such a pipelined ADC, each stage functions with a delay of 1 clock period, with respect to the previous stage. Also, each successive stage receives as its input the remainder that occurred from the earlier conversion. In this way, the first-stage functions as a relatively coarse resolution ADC and generates the first set of MSBs. This sequence is fed to the digital-to-analog converter (the m-bit DAC in Figure 11.16) and is subtracted from the original sequence before conveying to the next stage following buffer amplification. This process is shown as Amp and V_{res} in Figure 11.16. During the next clock period, the next m-bit stage stores the difference between the earlier output sequences and converts the result into a further bit sequence, and so on.

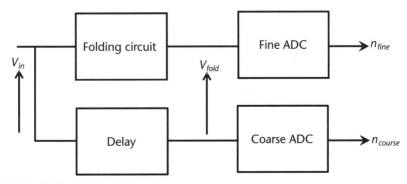

Figure 11.15 Folding ADC architecture.

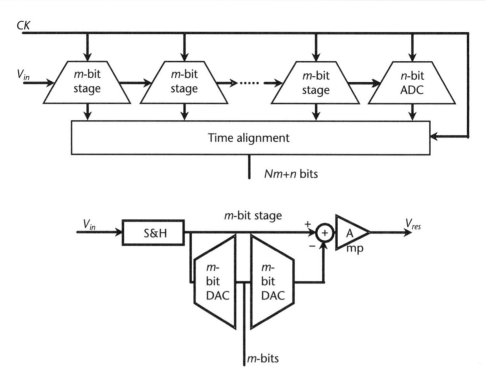

Figure 11.16 The overall architecture of a pipelined ADC. (© Artech House, 2016.)

The final output, from the N m-bit ADC stages, is $Nm+n$ bits. This type of ADC functions fairly rapidly, provides high resolution, and occupies a relatively small area of RFIC die.

Another type of ADC that is often used in lower-speed systems, such as instrumentation, is the successive-approximation ADC. However, as these types of ADC are generally unsuitable for high-speed RF conversion, they will not be covered here.

11.3.4 Time-Interleaved ADCs

These ADC architectures implement a range of m parallel sub-ADCs and each sub-ADC samples the incoming signal at every m'th cycle of the system clock. This arrangement causes the sample rate to be increased by a factor of m compared to the sampling rate capability of each individual sub-ADC. Time-interleaved ADCs are therefore usually very fast indeed.

The g'th sub-ADC configuration is shown in Figure 11.17.

The overall arrangement for the time-interleaved ADC is indicated in Figure 11.18.

Notice how each consecutive sub-ADC is clocked by successively delayed clocks, working downward:

- Sub-ADC1 is sampled by 3× clock delays.
- Sub-ADC2 is sampled by 2× clock delays.

11.3 ADC Architectures

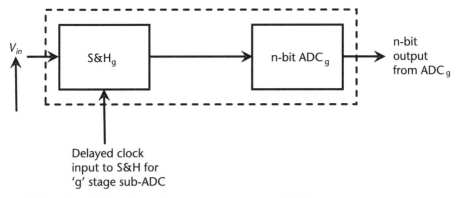

Figure 11.17 Sub-ADC configuration for a time-interleaved ADC.

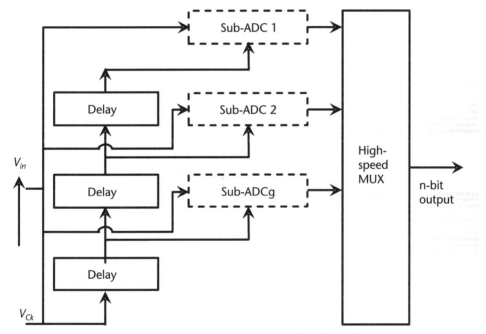

Figure 11.18 Overall arrangement for the time-interleaved ADC (TI ADC).

- Sub-ADC$_g$ is sampled by just one clock delay.

In a practical system, physical differences between the m sub-ADCs degrade the overall performance but provided the system is integrated (RFIC) such differences are generally negligible. Delays are of the order of nanoseconds. ENOBs are typically in the range 4 to 6 and TI ADCs can operate at frequencies well into the gigahertz range.

For further detail concerning time-interleaved ADCs, [9–11] should be consulted.

11.4 Digital-to-Analog Converters

A received digital signal is only useful directly where this signal is going to be sent to a computer or to a similar all-digital-processing system. Otherwise, where the end product is required to be sound, vision (or some mechanical movement) all this is analog and the incoming digital signal must be converted to analog. The subsystem that performs this function is termed a digital-to-analog converter (DAC).

11.4.1 Basic Structure and Functionality of a DAC

Figure 1.12 shows the basic schematic of a DAC. Figure 11.19 indicates this filter, as well as the sampling portion of the ADC and all the frequency components involved. The static output voltage versus bit code transfer function is shown in Figure 11.20.

After the reconstruction filter, the signal is a much closer replica of the original. In the time domain (i.e., as a waveform), it loses its former jagged and stepped edges and becomes rounded. Viewed in the frequency domain, its spectrum has far fewer unwanted high-frequency components.

11.4.2 DAC Resolution, Speed, and Figures of Merit

11.4.2.1 Resolution

Referring to Figure 11.20 for the minimum voltage step, this ratio is given by:

$$V_{LSB} = \frac{V_{ref}}{2^N} \tag{11.8}$$

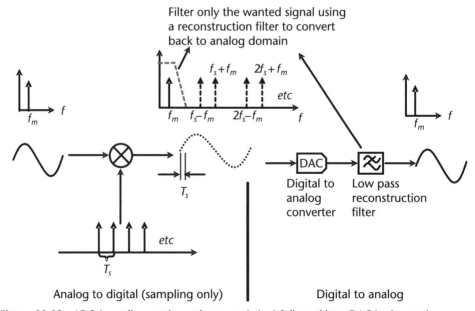

Figure 11.19 ADC (sampling section only, transmission) followed by a DAC in the receiver.

11.4 Digital-to-Analog Converters

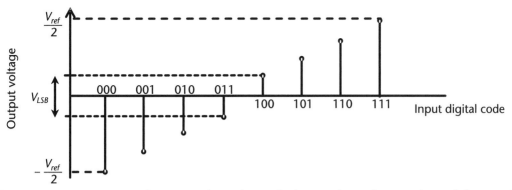

Figure 11.20 Static output voltage versus bit code transfer function for a 3-bit DAC. (From: [3]. © Artech House, 2016.)

where V_{ref} is the reference voltage.

However, it is necessary to bear in mind that some DAC architectures automatically cause the resolution to decrease when operating at high speed. This feature has to be checked and, when observed, allowed for in the system design.

Example 11.3

When the reference voltage in a DAC is 2V and the circuit is a 4-bit DAC, what is the voltage excursion for the LSB?

Using (11.8) and substituting these values, the answer is 0.125V or 125 mV.

As the number of bits accommodated rises, so the V_{LSB} value decreases, monotonically. For all other aspects considered, use as large a number of bits as possible.

11.4.2.2 Speed

The speed at which a DAC can function is defined as the maximum sampling rate and this corresponds to the maximum clock frequency at which the DAC's output remains what would be expected and also remains accurately close to the original transmission.

Another group of important characteristics that is next considered here starts with what can be termed quasi-static figures of merit. The term quasi-static refers to relatively slowly varying inputs.

11.4.2.3 Differential Nonlinearity

This is defined as a voltage error determined using the equation:

$$\text{error} = \frac{\delta V - V_{LSB}}{V_{LSB}} \quad (11.9)$$

in which δV is the small difference between the output voltage detected between two successive codes and V_{LSB} can be calculated using (11.8).

11.4.2.4 Integral Nonlinearity

This is defined as the error comparing an ideal linear output voltage transfer function and the actual detected version.

11.4.2.5 Offset

This is the difference between the theoretical values of voltages and currents in the circuit compared with the actual (practical) values. The differences are caused by variations in the physical parameters associated with the semiconductor components.

Integral nonlinearity and offset are illustrated in Figure 11.21.

Integral nonlinearity results in output voltage waveform shape discrepancy between the ideal and the practical. Contrastingly, offset (shown as output voltage offset in Figure 11.21) refers to the difference in ideal versus practical voltage or current levels.

11.4.2.6 Dynamic Figures of Merit

Four dynamic figures of merit are cited here. Each is concerned with output power (P_{out}) considerations.

1. *Spurious-free dynamic range:* This is defined as the difference between P_{out} in the signal channel and the highest power level of any spurious (unwanted) frequency component.
2. *Signal-to-noise and distortion ratio (SINAD):* This is defined as the difference between P_{out} of the desired signal and the combined noise and spurious power levels.
3. *Harmonic distortion:* The effects of harmonic distortion are also very important characteristics associated with DACs. Harmonic distortion, in particular, leads to undesired spectral components, some of which may be impossible to filter out. These effects also limit the dynamic range, that is, the ratio (usually in decibels) between the maximum and minimum output

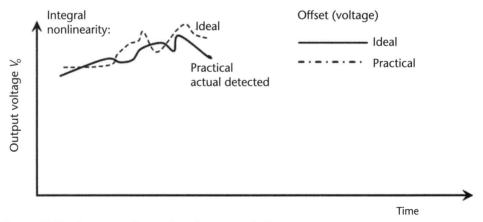

Figure 11.21 Concepts of integral nonlinearity and offset.

voltages. Defining harmonic distortion requires taking the aggregate sum of the power in all the harmonics that are present at the output.

4. *Adjacent channel power ratio (or CIMR3):* Adjacent channel power ratio is defined in Section 11.2.10. The abbreviation CIMR3 refers to the carrier-to-(third-order)-intermodulation ratio, which is defined and discussed in Chapter 10.

11.4.3 Some Practical Aspects of High-Speed DACs

Architectures for digital radio systems operating at high frequencies demand special technological approaches. Millimeter-wave systems, for example, require sampling at typically several gigasamples per second (Gsps). Hybrid technologies are preferred for DACs in such situations. A good example is the segmented DAC which has a twofold architecture designed as follows: a thermometer-coded sub-DAC processing the MSBs and a binary-weighted sub-DAC handling the LSBs.

Sampling rates associated with these types of DACs can reach several Gsps, which renders them suitable for high-microwave and millimeter-wave applications.

References

[1] http://www.gaussianwaves.com/2011/07/sampling-theorem-baseband-sampling/.

[2] www.ws.binghamton.edu/fowler/fowler%20personal%20page/EE521_files/II-2%20BP%20Sampling_2007.pdf.

[3] Camarchia, V., R. Quaglia, and M. Pirola, *Electronics for Microwave Backhaul*, Norwood, MA: Artech House, 2016.

[4] Gersho, A., and R. M. Gray, *Vector Quantization and Signal Compression*, New York: Springer, 1991.

[5] Neu, T., "Clocking the RF ADC with Phase Noise Instead of Jitter," *Microwave Journal*, August 2017, pp. 100–106.

[6] Analog Devices, *The Data Conversion Handbook*, Norwood, MA, Newnes, 2005.

[7] Stanley, W. D., *Operational Amplifiers with Linear Integrated Circuits*, Merrill's International Series in Electrical and Electronics Technology, C. E. Merrill, 1984.

[8] https://enwikipedia.org/wiki/Flash_ADC.

[9] Vogel, C., and H. Johansson, "Time-Interleaved Analog-to-Digital Converters: Status and Future Directions," *IEEE ISCAS 2006*, 2006, pp. 3386–3389.

[10] El-Chammas, M., "The World of Time-Interleaved ADCs: From Theory to Design," Texas Instruments, Inc., June 17, 2012.

[11] Duan, Y., "Design Techniques for Ultra-High-Speed Time-Interleaved Analog-to-Digital Converters (ADCs)," Technical Report No. UCB/EECS-2017-10, May 1, 2017, http://www2.eecs.berkeley.edu/Pubs/TechRpts/2017/EECS-2017-10.html.

CHAPTER 12

Radio Frequency Sources

12.1 Some Fundamental Aspects of RF Oscillators

An oscillator is typically specified according to the following characteristics in terms of its output:

- Center frequency;
- Frequency stability (e.g., drift with temperature);
- Levels and frequencies of spurs or other unwanted output components;
- Phase noise.

Any oscillator requires three elements:

1. A source of energy (usually the DC supply);
2. A device that will provide signal gain (usually a transistor);
3. A means of defining the frequency of oscillation (a tuned circuit, microwave cavity, or quartz crystal).

This overall situation is exhibited in Figure 12.1.

In general the output (or center) frequencies of RF oscillators can range from some tens of megahertz up to maybe 100 GHz or more. Crystal-based reference oscillators operate at some tens of megahertz through to 100 or 125 MHz. Other RF oscillators such as the traditional Colpitts or Hartley types (also dielectric resonator oscillators) can be designed to function at frequencies well into the microwave bands. VCOs can be designed for very high frequency (VHF) through extrahigh frequency (EHF) applications.

Maintaining frequency stability and minimizing the levels of unwanted output components and phase noise all represent important aspects of oscillator design and choice. Phase noise is sufficiently significant to warrant an entire section in this chapter dealing exclusively with this aspect.

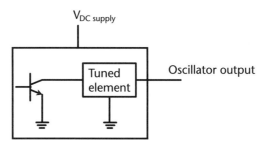

Figure 12.1 Basic requirements for an oscillator.

It is always vital to have a highly stable, low-noise reference frequency source and this requirement is almost always met by implementing an oscillator based on a quartz crystal as the frequency-determining element. This is addressed in the next section.

12.2 Quartz Crystal Oscillators

12.2.1 The Quartz Crystal

Crystalline quartz is one of the most enduring materials still used in many electrical and electronic systems today. The piezoelectric effect offered by quartz makes it particularly significant in many circuits, including high-stability oscillators.

The phenomenon of piezoelectricity was discovered in 1880 by the French physicists Jacques and Pierre Curie. Without a doubt, this must be about the most exceptional historical example of a phenomenon discovered during the century before last but that remains important in today's electronics.

Piezoelectricity is the phenomenon, available only in certain specific materials, in which the application of mechanical pressure results in an electrical output. This is why piezoelectric transducers are so frequently used as the transducer that responds to the small surface ripples (recorded sounds) at the end of record players for vinyl disks. The piezoelectric effect is also reciprocal, that is, electrical oscillations applied to a piezoelectric crystal such as quartz slightly alter its mechanical dimensions, hence varying its capacitance. For modern practical applications, the crystalline quartz is almost always synthetically manufactured and is available encapsulated in a rugged assembly. This assembly comprises a small rectangular cut of single-crystal quartz held between metal electrodes. Various angles of cuts through a quartz crystal yield differing properties. The equivalent electrical circuit associated with a quartz crystal is shown in Figure 12.2.

In the equivalent electrical circuit of Figure 12.2, the elements R, L, C all have their origins in the fundamental motional character of the crystal resonator. This series R, L, C arm is shunted by the parallel (package) capacitance C_p. The circuit configuration results in a fairly complex reactance trend as a function of frequency, exhibiting two distinct resonances as indicated in Figure 12.3.

The series and parallel resonant frequencies are given by (12.1) and (12.2), respectively:

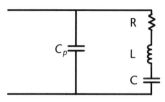

Figure 12.2 Equivalent electrical circuit for a quartz crystal resonator.

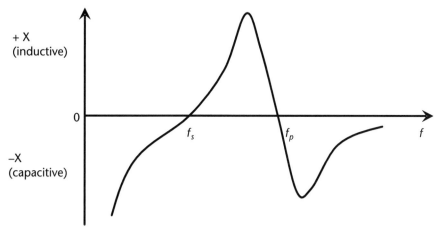

Figure 12.3 Reactance of a crystal resonator plotted against frequency.

$$f_s = \frac{1}{2\pi\sqrt{LC}} \qquad (12.1)$$

and

$$f_p = \frac{1}{2\pi\sqrt{L\left(\dfrac{C_p C}{C_p + C}\right)}} \qquad (12.2)$$

Due to the extremely low losses associated with quartz crystal resonators, the unloaded Q-factors tend to be very high (typically 100,000 or so). Drift of frequency as a function of ambient temperature is markedly low, being less than 0.001% per degrees Celsius. It is also possible to further stabilize a quartz crystal oscillator by placing the crystal assembly within a small temperature-controlled oven. This is usually a practical proposition for implementation in an outdoor (or relatively large indoor) installation but not for a small mobile system.

12.2.2 Quartz Crystal-Based Oscillators

Because the circuit is inductive at frequencies between the series and parallel resonant values (Figure 12.3), the quartz crystal is usually connected so as to replace the inductor in either a Colpitts or a Pierce oscillator. These general types of oscillators are well covered in the literature and an example of a Pierce crystal oscillator circuit is shown in Figure 12.4.

In the circuit shown in Figure 12.4, resistors R_{B1}, R_{B2}, and R_E all determine the DC bias conditions for the transistor. Inductor L_s is the RF choke preventing significant RF energy from reaching the DC supply. Bypass capacitors C_{SB} and C_{EG} direct RF signal current away from the DC biasing components.

The quartz crystal, operating inductively and shunting capacitors C_1 and C_2, determines the oscillation frequency. Quartz-crystal-based oscillators are essential circuits for the generation of the highly stable source frequencies needed in communications systems. Actual frequencies of oscillation vary from some tens of megahertz up to typically 100 MHz or 125 MHz. Even where a final system frequency requirement may be many tens of gigahertz, a quartz crystal reference oscillator is almost invariably used.

Quartz crystal oscillators continue to be favoured for highly stable, spectrally pure frequency sources. These types of oscillators continue to hold-off the challenges mounted by the following technologies: surface-acoustic wave (SAW) tuning elements, and microelectromechanical system (MEMS) tuning elements.

This oscillator exhibits a phase noise of around −140 dBc/Hz, 100 Hz away from the carrier. This is broadly in line with expectations (see Section 12.6). Frequency synthesizers, for example, described later in this chapter, require quartz crystal reference oscillators.

12.3 Oscillators Controlled by Dielectric Resonators

In contrast to crystal oscillators, dielectric resonator oscillators (DROs) operate directly at microwave frequencies. The main trade-offs are reduced stability and a somewhat poorer phase noise performance. Where these trade-offs can be tolerated,

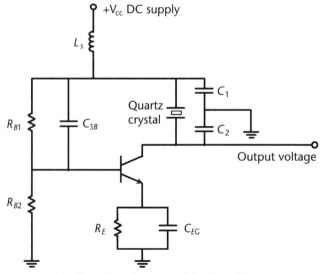

Figure 12.4 Quartz-crystal oscillator based on Pierce's basic configuration.

12.3 Oscillators Controlled by Dielectric Resonators

DROs provide an economically viable frequency source in many communications systems.

Dielectric resonators are introduced in Chapter 6 (Section 6.9) in the context of designing bandpass filters. These unique passive components can also be used as the frequency-determining element in oscillators: DROs. Figure 12.5 is a schematic diagram showing a dielectric resonator (DR) coupled to a microstrip line and thereby forming a basis toward a DRO.

As might be anticipated intuitively, the smaller the coupling distance between the microstrip and the DR the tighter will be the degree of coupling. An equivalent circuit for this physical situation comprises a parallel tuned (or tank) circuit representing the DR and coupled to the microstrip line by means of a theoretical transformer having turns ratio N:1. This equivalent circuit is shown in Figure 12.6.

The reflection coefficient, seen on the terminated microstrip line looking toward the DR, is given by the following equation.

$$\Gamma = \frac{N^2 R}{2Z_0 + N^2 R} \tag{12.3}$$

where N is the turns ratio of the theoretical transformer, R is the equivalent parallel resistance of the dielectric resonator, and Z_0 is the characteristic impedance of the microstrip.

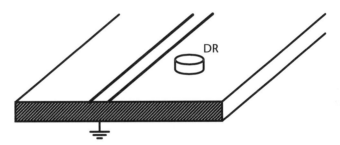

Figure 12.5 DR coupled to an adjacent microstrip line.

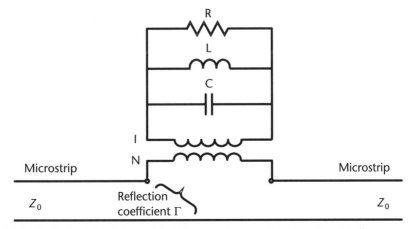

Figure 12.6 Equivalent circuit for a dielectric resonator coupled to a microstrip line.

Example 12.1

Calculate the reflection coefficient associated with a dielectric resonator coupled to a 50-Ω microstrip line where the following parameters apply: resonant frequency 16 GHz, the measured equivalent capacitance $C = 4$ pF, and the Q-factor = 2,000.

- *Step 1:* Using the fundamental expression for Q-factor: $Q = 2\pi fCR$

$$2{,}000 = 2\pi \times 16 \times 10^9 \times 4 \times 10^{-12} \times R$$

which leads to the result $R = 4{,}974\Omega$.

Initially assuming that the turns ratio N is 5 and then substituting this value and the above value of R into (12.3) give the reflection coefficient as:

$$\Gamma = \frac{5^2 \times 4.974 \times 10^3}{2 \times 50 + 5^2 \times 4.974 \times 10^3} \sim 1$$

that is, very nearly total power reflection.

For any DR the Q-factor will always be high; therefore, it would be necessary for the turns ratio to be set much lower than 5 and likely reversed (i.e., a fraction) for the reflection coefficient to become smaller than 1.

The above outline treatment is fundamental for the design of DROs.

For example, an 8.5-GHz DRO exhibits a phase noise of around −120 dBc/Hz, 100 Hz away from the carrier, which is fairly typical for DROs. This phase noise is somewhat worse than the crystal oscillator but it applies at a much higher frequency and is in line with general expectations (see Section 12.6).

12.4 VCOs

In many instances within communications systems, it is necessary to automatically (dynamically) control the frequency of an oscillator. The requirement is to provide this frequency variation very rapidly, that is, in times amounting to microseconds or even nanoseconds and the concept is illustrated in Figure 1.13.

The terminology used to define this type of oscillator is VCO or occasionally voltage-tuned oscillator (VTO). The basic component most frequently employed for this task is the varactor diode, which is described in Section 2.3.4 of Chapter 2. Although the capacitance-voltage characteristic is highly nonlinear, it is almost always sufficient for this variation to effectively control the resonant frequency of the oscillator's tank circuit and hence the output frequency. A typical VCO circuit employing a bipolar transistor is shown in Figure 12.7. This device could be a GaAs HBT as described in Chapter 2 (Section 2.4.3), which would provide for a relatively low-noise contribution.

All the resistors in Figure 12.7 provide appropriate DC biasing for the transistor. Capacitors C_{BL1}, C_{BL2}, and C_{BL3} all provide DC blocking functions. Capacitors C_1 and C_2 operate in conjunction with inductor L and the varactor diode to tune the circuit.

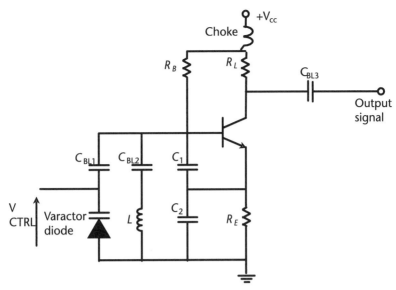

Figure 12.7 VCO implementing a bipolar transistor (BJT).

VCOs can typically be tuned over 500-MHz bandwidths, for example, from 10.5 to 11 GHz. Associated phase noise is around −82 dBc/Hz at a frequency shifted 10 kHz from the carrier. This means that when calculated or measured 100 Hz from the carrier the phase noise will definitely be worse, perhaps as high as −70 dBc/Hz. Phase noise as a general challenge is now treated in some detail.

12.5 Importance and Impact of Phase Noise

As outlined in Chapters 2 and 9, all electronics tends to be negatively impacted by noise, and RF oscillators are no exception. The main issue here is phase noise, which is described in Chapter 9.

From the viewpoint of phase (and thermal) noise, it is only necessary to consider the output noise power as a function of frequency, taking into account this noise power from the carrier frequency upward. This trend is shown in Figure 12.8.

The challenge is to mathematically determine all the required phase and thermal noise components. Toward this, consider the fact that the noise behavior is due to random phase perturbations of the oscillator output (jitter in the time domain). This behavior, in turn, originates from imperfections in the materials used in any oscillator: the resonating components (L, C, quartz crystal) and the transistor. It is useful to begin the analysis by expressing the frequency components as shown in (12.4), where it should be borne in mind that it is usually best mathematically to start with a cosine function rather than a sine function:

$$v(t) = A\cos\{2\pi f_c t + P_m \sin(2\pi f_m t)\} \qquad (12.4)$$

In which the quantity P_m represents the amplitude of the phase noise.

This expression expands to involve a product of two sine functions, as well as a cosine wave so that (assuming $P_m \ll 1$) the next expressions become:

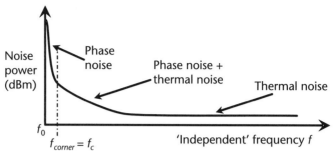

Figure 12.8 Oscillator output noise power versus frequency.

$$v(t) = A\cos(2\pi f_c t) - A\sin(2\pi f_c t) P_m \sin(2\pi f_m t)$$

the sine product term of which expands to form two adding cosine terms, making a total of three terms as follows:

$$v(t) = A\cos(2\pi f_c t) - \frac{AP_m}{2}\cos\{2\pi(f_c - f_m)t\} + \frac{AP_m}{2}\cos\{2\pi(f_c + f_m)t\} \quad (12.5)$$

This equation clearly shows the ideal center frequency component together with a lower side-frequency component $(f_c - f_m)$ and an upper side-frequency component $(f_c + f_m)$. In practice, phase noise is spread over a wide range of frequencies so that instead of the single components indicated in (12.5), there is a wide spread of side frequencies all with amplitude P_m as a function of f_m. The result is usually expressed as decibels with respect to the carrier amplitude (i.e., dBc).

The remainder of this section is focused toward presenting a modified version of Leeson's formula applying to single-sideband noise spectral density. Leeson's formula is fairly widely used aiding the understanding of phase noise.

A useful representation leading to an analysis of phase noise is provided by the feedback amplifier shown in Figure 12.9.

Here it is assumed that the overall gain of the complete feedback amplifier is 1.0, the center (resonant) frequency is f_0, and the loaded Q-factor of the resonator is Q_L. A thermal noise component at the input, having frequency $(f_m - f_0)$ (also its

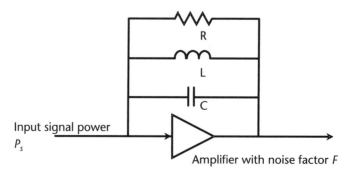

Figure 12.9 Conceptual diagram of an amplifier with resonator feedback (phase noise interpretation).

correlated counterpart [$f_m + f_0$]) introduces instantaneous phase fluctuations at the output. Thermal noise, having average power $FkTB$, is present at all frequencies in the range and therefore operates uniformly (F being the noise factor).

However, a further very important noise source is also actively present: flicker noise (or $1/f$ noise). This phenomenon is discussed in Chapters 2 and 9. The corner frequency f_c associated with flicker noise (see Figure 12.8) marks the frequency at which the dominance of flicker noise gives way to increasingly significant thermal noise as frequency increases away from the carrier.

This basic noise model corresponds to a phase modulator connected to the input of the unity-gain feedback amplifier, with its spectrum normalized to the input signal power P_s:

$$S_\varphi(f_m) = \frac{FkT}{P_s}\left(1 + \frac{f_c}{f_m}\right) \tag{12.6}$$

Using this expression, the unilateral phase noise spectral density as a function of the offset frequency f_m can be obtained [3, 4] (normalized to the signal power).

In this analysis, it is assumed the oscillator is a VCO (see Section 12.5), tuned with a varactor diode that activates a frequency variation per voltage change $\frac{\Delta f}{\Delta v} = K_0$ and for which R is the varactor's equivalent noise resistance. This noise resistance can be as low as 50Ω but can extend to several tens of kilo-ohms. The unilateral phase noise power spectral density as a function of the offset frequency f_m is the modified Leeson's formula [2]:

$$L(f_m) = \frac{FkT}{2P_s}\left(1 + \frac{f_c}{f_m} + \frac{f_0^2}{4Q_L^2 f_m^2} + \frac{f_0^2 f_c}{4f_m^3}\right) + \frac{2kTRK_0^2}{f_m^2} \tag{12.7}$$

In this equation, F is the noise factor, k is Boltzmann's constant, and T the is absolute temperature. The coefficient $FkT/2P_s$ effectively scales the first four terms through the choice of F (linear factor) and P_s. Choosing $F = 2$ and $P_s = 1\mu W$ usually results in a well-scaled formula. Expanding the right side of (12.7) yields five terms—a, b, c, d, and e—and it is important to appreciate the meaning and implications of each term:

- a: Thermal phase noise $FkT/2P_s$, which is independent of frequency;
- b: Flicker-induced phase noise, which is inversely proportional to f_m;
- c: Upconverted thermal FM noise (sometimes referred to as random-walk phase noise), which is inversely proportional to f_m^2;
- d: Upconverted flicker FM noise, which is inversely proportional to f_m^3;
- e: Amplitude-to-phase modulation (AM to PM) resulting from the varactor tuning diode's loading effect.

The phase noise power spectral density, at frequency $(f_0 + f_m)$ above the center frequency f_0, has units dBc/Hz. This determines the noise power, with respect to the desired signal power, in a 1-Hz bandwidth centered within $(f_0 + f_m)$.

Equation (12.7) expresses all the frequency-dependent trends for phase noise, as frequencies extend upwards away from the carrier or center frequency f_0. This equation can also be used to calculate the theoretical value of the phase noise in dBc/Hz at the offset frequency $(f_0 + f_m)$ above the carrier. It is particularly significant to evaluate the phase noise at an offset frequency very close to the carrier because this is in the vicinity of the worst-case phase noise.

Example 12.2

Consider a quartz-crystal-based oscillator for which the following parameters apply: noise factor $F = 2$, $T = 290$ (K), corner frequency $f_c = 30$ MHz, oscillation (center) frequency $f_0 = 125$ MHz, loaded Q-factor of the resonant circuit = 1,000, noise equivalent resistance $R = 1$ kΩ, conversion coefficient $K_0 = 1$ MHz/V. k (Boltzmann's constant) = 1.38×10^{-23} J. K^{-1}.

Calculate the phase noise at the output of this oscillator, at a frequency 100-Hz away from the center frequency.

- *Step 1:* Calculate the coefficient $FkT/2P_s$ with the reference power P_s set at 1 μW;

$$\frac{FkT}{2P_s} = \frac{2 \times 1.38 \times 10^{-23} \times 2.9 \times 10^2}{2 \times 10^{-6}} = 4 \times 10^{-15} \quad (12.8)$$

- *Step 2:* Referring to the bracketed terms beneath (12.7), term a is the same result as (12.8).
- *Step 3:* The spot frequency 100 Hz away from the center frequency f_0 is actually $(f_0 + f_m)$, which = 125 + 0.0001 MHz = 125.0001 MHz in this instance. This value must be substituted for f_m in (12.7). Here this spot frequency is very nearly equal to f_0, which means that all the appearances of the ratios:

$$\frac{f_0}{f_m} = \frac{f_0^2}{f_m^2} \sim 1 \text{ (very closely indeed)}$$

- *Step 4:* Calculate term (b):

$$b = a\frac{f_c}{f_m} = 4 \times 10^{-15} \times \frac{3 \times 10^7}{125 \times 10^6}$$

Giving: $b = 0.96 \times 10^{-15}$

- *Step 5:* Calculate term (c):

12.5 Importance and Impact of Phase Noise

$$c = a\frac{f_0^2}{4Q_L^2 f_m^2} = a\frac{1}{4Q_L^2} = 4\times10^{-15} \times \frac{1}{4\times10^6}$$

Giving: $c = 1 \times 10^{-21}$

- *Step 6:* Calculate term (d);

$$d = a\frac{f_0^2 f_c}{4f_m^3} = a\frac{f_c}{4f_m} = 4\times10^{-15} \times \frac{3\times10^7}{4\times125\times10^6}$$

Giving: $d = 2.4 \times 10^{-16}$

- *Step 7:* Term (e) outside the bracket in (12.7) is also required:

$$e = \frac{2kTRK_0^2}{f_m^2} = \frac{2\times1.38\times10^{-23} \times 2.9\times10^2 \times 10^3 \times 10^{12}}{125.0001^2 \times 10^{12}}$$

Giving: $e = 5.13 \times 10^{-21}$

- *Step 8:* Add all five terms to give the overall (linear) phase noise level:

Overall $= a + b + c + d + e$

$$= 4\times10^{-15} + 0.96\times10^{-15} + 1\times10^{-21} + 2.4\times10^{-16} + 5.13\times10^{-21}$$

which is approximately 5.2×10^{-15}

Taking logs, the value is −14.28, and multiplying by 10 gives the phase noise as: phase noise at 100-Hz offset = −142.8 dBc/Hz.

This value is somewhat lower than typically found for commercially available quartz-crystal oscillators operating at this output frequency, which is around −130 dBc/Hz. However, careful research and development have shown that this value can be improved to −144 dBc/Hz [3, 4].

An example of a typical phase noise versus frequency away from the carrier curve is shown in Figure 12.10 (for a nominally 125-MHz crystal oscillator).

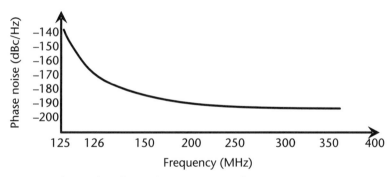

Figure 12.10 Typical crystal oscillator phase noise versus frequency curve.

In Figure 12.10, the first frequency, labeled 125, is actually 125.0001 MHz (i.e., shifted away from the 125-MHz carrier by just 100 Hz). It should be clear at this point that for any oscillator (or other frequency source) the phase noise always worsens the closer to the carrier the spot frequency gets. In practice, the 100-Hz shift is almost always the value chosen for high-performance reference oscillator.

The classic danger for systems designers choosing a suitable oscillator is to simply accept the manufacturers' data regarding phase noise while failing to ask for further information. Often, phase noise is specified at, say, 1 kHz away from the carrier, where the phase noise may be as low as −150 or even −160 dBc, or even 1 MHz away from the carrier, yielding perhaps −180 dBc/Hz.

It is almost always essential to know the worst news (i.e., the value of the phase noise close-in to the carrier, usually 100 Hz away). It is necessary to return to the subject of phase noise where frequency synthesizers are considered later in this chapter.

12.6 Frequency Multipliers

It is often impractical to attempt the direct generation of extremely high-frequency (microwave or millimeter-wave) signals. Reasons include the design compromises that would be required to yield a sufficiently stable, low-noise, and spectrally pure frequency source. Instead, a relatively low-frequency signal source can be used to drive a frequency multiplier circuit, and the required harmonic can then be selected for the much higher output frequency.

A typical diode-based frequency multiplier circuit is shown in Figure 12.11.

In Figure 12.11, v_{in} is the (lower frequency) input signal, D is the semiconductor diode, and the parallel RLC arrangement is the tuned (or tank) circuit. The most important feature to appreciate is that the diode I/V characteristic is highly nonlinear, as covered in Chapter 2. The diode I/V expression is repeated here:

$$I_F = I_{SAT}\left(e^{\alpha V} - 1\right) \tag{12.9}$$

All the quantities in this equation were defined in Chapter 2.

Assuming the (nonlinear) resistance of the diode is R_D, then the output voltage v_{out} developed across the tuned circuit at resonance will be, according to the voltage-divider principle:

$$v_{out} = \frac{v_{in} R}{R + R_D} \tag{12.10}$$

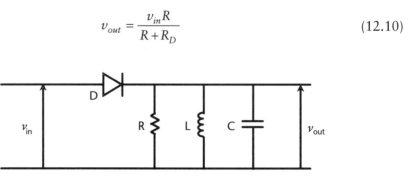

Figure 12.11 Basic diode-based frequency multiplier circuit.

However, due to the nonlinear behavior of the diode, the output voltage can also be expressed as a series of terms in a Taylor expansion:

$$v_{out} = a_0 + a_1 v_{in} + a_2 v_{in}^2 + a_3 v_{in}^3 \ldots \quad (12.11)$$

The aim is to determine v_{out} as a function of the input frequency f_i in order to establish the capability of this circuit to produce multiples of f_i at the output. The input signal will be sinusoidal because it comes from (typically) a crystal oscillator or a DRO. Therefore, this input signal can be expressed as $v_{in} = I\sin(2\pi f_i t)$ and substituting this expression into (12.11) results in:

$$v_{out} = a_0 + a_1 I \sin(2\pi f_i t) + a_2 I^2 \sin^2(2\pi f_i t) + a_3 I^3 \sin^3(2\pi f_i t) + a_4 I^4 \sin^4(2\pi f_i t) + \ldots$$

All the power terms in this expression can be further expanded with the following result:

$$\begin{aligned} v_{out} = {} & a_0 + a_1 I \sin(2\pi f_i t) + \frac{a_2 I^2}{2} - \frac{a_2 I^2}{2}\cos(4\pi f_i t) \\ & + \frac{a_3 I^3}{2}\{3\sin(2\pi f_i t) - \sin(6\pi f_i t)\} + \frac{3a_4 I^4}{8} \\ & + \frac{3a_4 I^4}{8}\{\cos(8\pi f_i t) - 4\cos(4\pi f_i t)\} + \ldots \end{aligned} \quad (12.12)$$

This expression can be simplified further, but it is already evident from (12.12) that harmonics up to at least the eighth are available at the output. The amplitudes of these harmonics depend on the maximum amplitude of the input signal (I) as well as the coefficients (a_n) associated with the nonlinear diode characteristics. The resonant frequency of the tuned circuit can be chosen to select the required output frequency. For example, if an output frequency of 84 GHz was required and the eighth harmonic was selected, then the input frequency has to be 10.5 GHz, which could readily be provided by, say, a DRO.

Some frequency multipliers implement transistors rather than the diode selected in this example. However, making the circuitry more complex has the serious disadvantage of adding to the noise in the final output.

12.7 Frequency Dividers

Occasionally it is necessary to divide-down a particular frequency and various options are available for this purpose. A good example of the requirement is within a PLL-based frequency synthesizer, the subject of the next section.

The main choice is between analog and digital technologies and in the example shown here (Figure 12.12) an analog configuration is indicated.

The operation of this arrangement is as follows.

The signal f_{in} required to be halved enters one port of the mixer while a feedback signal $1/2 f_{in}$ enters the other input port. The mixer combines these two signals such that it produces two output frequency components, $1/2 f_{in}$ and $2/3 f_{in}$, which

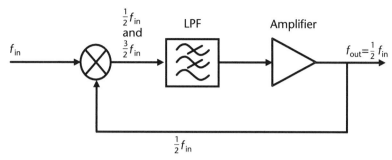

Figure 12.12 Frequency halving arrangement using analog circuitry.

are the two signal components fed to the LPF. Next, the lowpass filter (LPF) removes the higher-frequency component, that is, $3/2 f_{in}$ and the output is $1/2 f_{in}$ as required. It is almost always necessary to amplify the output of the LPF using a low-noise amplifier (Chapter 9 covers LNA design).

12.8 Phase-Locked-Loop-Based Frequency Synthesizers

Several references include excellent coverage of frequency synthesizers in general notably Pozar [1] and Chenakin [5], including phase locked loop (PLL) types. PLLs are classified according to whether they can detect and process signals by: phase alone, frequency alone, or both phase and frequency.

12.8.1 Basic Configuration

- The output is divided by N so this signal applied to the phase detector closely matches f_0.
- The phase detector generates a small voltage proportional to the difference in the phases of the signals f_0 and f (ideally this voltage tends toward zero).
- This small voltage is amplified and lowpass-filtered before applying the result to make small corrections to the frequency of the VCO so as to align the phase of the VCO output with that of the reference source.
- Hence, the final output ($N f_0$) has spectral characteristics very similar to those of the reference source, except at a much higher frequency.
- Among these characteristics, the phase noise behavior is very similar to that of the reference source.
- $N f_0$ is always very much higher than f_0 (i.e., $N f_0 \gg f_0$).

12.8.2 The Fractional-N Frequency Synthesiser

The fractional-N type of PLL-based frequency synthesizer represents an important variation on the basic PLL-based approach described above.

In this configuration, the frequency divider is designed such that $(N + 1)$ is enabled as well as simply N.

12.8 Phase-Locked-Loop-Based Frequency Synthesizers

Also, a second (digital) control loop is added between the phase detector and the frequency divider. The purpose of this additional loop is to enable the fractional-N control of the frequency divider. The extra elements are shown in Figure 12.13 (the remainder of the PLL synthesizer is the same as the one in Figure 1.15).

The subsystem shown in Figure 12.13 can operate to produce an output frequency that is an integer-plus-fractional multiple of the reference oscillator frequency. This is a very important attribute for a dynamic frequency source and the basis of operation is now described.

- Divide by N for a certain specific number of cycles of the output frequency f_{out}.
- Also, divide by $(N+1)$ for another, different, number of cycles.
- This causes the frequency divider to switch between division ratios of N and $(N+1)$ under the control of an adder that counts the number of cycles.

For example, if the VCO output is divided by $(N+1)$ every M cycles, and also divided by N for all intervening cycles, then the average output frequency is given by:

$$f_{out} = \left(N + \frac{1}{M}\right) f_0 \qquad (12.13)$$

Example 12.3

Suppose the reference oscillator frequency is 100 MHz and an output frequency of 9.428 GHz is required. This means that the overall division ratio is 94.28. The divisor N takes care of the integer portion (i.e., $N = 94$).

The additional (fractional) portion $(1/M)$ takes care of the fractional portion (i.e., 0.28), which means that $1/M = 0.28$ (i.e., $M = 3.5714$).

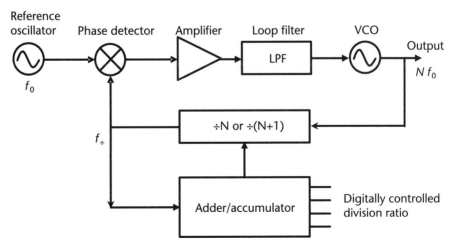

Figure 12.13 Configuration of a fractional-N frequency synthesiser.

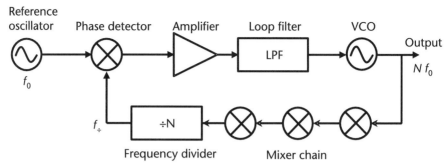

Figure 12.14 PLL-based frequency synthesizer including a mixer chain.

Now the VCO output is divided by $\{(N+1) = 95\}$ every $M = 3.5714...$ cycles and divided by 94 otherwise. It is better, for the $(N+1)$ case, to divide by 95 on every $1,000/3.5714 = 280$ cycle occasions during every sequence of 1,000 cycles and to divide by 94 in all other instances. In practice, the accumulator and frequency divider take care of this by digital processing.

Chenakin [5] describes details of various techniques for, in particular, greatly reducing the frequency offset in a PLL-based synthesizer. One interesting example, originally a patent due to Chenakin, is to introduce a mixer chain within the feedback path of a synthesizer as shown in Figure 12.14.

In this configuration, the mixer intermodulation products are readily filtered out by the loop filter.

Frequency synthesizers based on yttrium-iron-garnet (YIG) are occasionally used (YIG is also mentioned in Chapter 6). Such synthesizers can cover octave bandwidths such as 8 to 16 GHz, but these are uncommon in most commercial communications systems.

References

[1] Pozar, D. M., *Microwave and RF Design of Wireless Systems*, New York: John Wiley & Sons, 2001.

[2] Camarchia, V., R. Quaglia, and M. Pirola, *Electronics for Microwave Backhaul*, Norwood, MA: Artech House, 2016.

[3] Apte, A., et al., "Optimizing Phase-Noise Performance," *IEEE Microwave Magazine*, June 2017, pp. 108–123.

[4] Browne, J., "Putting a Lock on Crystal Oscillator Phase Noise," *Microwaves & RF*, July 2017, p. 26.

[5] Chenakin, A., "Frequency Synthesis: Current Status and Future Projections," *Microwave Journal*, April 2017, pp. 22–36.

CHAPTER 13

Frequency-Band Conversion

13.1 Introduction

There are many situations in communications systems (especially receivers) where signal frequency bands must be converted. When either downconversion or upconversion is applied, it is vital that all the information contained in the original (RF) frequency band be retained. In the case of downconversion, the new band of frequencies becomes centered on a much lower frequency usually known as the intermediate frequency (IF).

Downconversion is the most important process since this is used in many receivers to provide an IF that can readily be processed by the AGC (Chapter 8) followed with ADC and further digital circuitry. ADC and DAC are the subjects of Chapter 11. An outline of the first stages of this processing is provided by Figure 13.1 (omitting the AGC circuitry; see Chapter 8).

Communications systems increasingly operate using RF bands in the K-band region and higher. These bands are well beyond the reach of current (even prospective) ADCs, and therefore downconversion is essential. The somewhat idealized spectrum shown in Figure 13.2 illustrates this requirement. Many texts provide extensive information on frequency-band conversion [1–3].

While the full bandwidths are always required in practice, much of this chapter focuses on the center frequencies (i.e., the center RF and the center IF) together with various examples and relationships that are important.

The main requirements of a mixer are to achieve:

- Frequency conversion;
- Rejection of the image frequency;
- As low a conversion loss as possible;
- As low as possible noise figure (NF);
- As low as possible intermodulation distortion (IMD) (typically, P_3 values for mixers range from 15 dBm to 30 dBm);
- As high as possible isolation between RF and LO ports (this depends on the type of coupler, but it typically ranges between 20 and 40 dB).

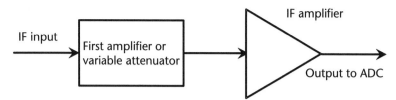

Figure 13.1 Post-IF signal processing.

Figure 13.2 Spectrum of RF and IF, including a numerical example.

Several mixer technologies are studied, but the Gilbert cell mixer is of such significance that it receives particularly detailed treatment.

13.2 Fundamentals of Mixers

13.2.1 Basic Features

RF mixers implement semiconductor devices, either diodes or transistors, and the fundamentals of these key devices are covered in Chapter 2. Diodes or transistors are selected because their inherently nonlinear I/V characteristics are essential for mixers. However, these devices must be designed into three-port circuits because two inputs are always required: the RF signal and a local oscillator (suitable oscillators are described in Chapter 12). The output is the IF as described above and the general configuration of a downconverting mixer is shown in Figure 13.3.

It will be the technology options for the central block in Figure 13.3, the mixer device and its circuit, that form the main subject matter of this chapter. The circle

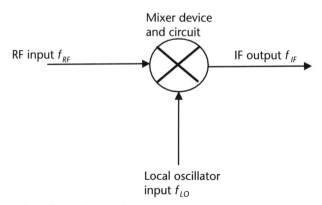

Figure 13.3 General configuration of a downconverting mixer.

with an interior cross is the accepted systems-level symbol for mixers and some other related circuits such as phase detectors.

In the configuration of Figure 13.3, the following frequency condition applies: Both f_{RF} and $f_{LO} \gg f_{IF}$

For example, taking the data shown in Figure 13.2, it is clear that:

$$f_{RF} (= 29 GHz) \gg f_{IF} (= 1.9 \text{ GHz})$$

It is necessary for the mixer devices, diodes or transistors, to be driven strongly by the RF and local oscillator inputs in order to take maximum advantage of the nonlinear behavior.

For the downconverting mixer shown (most are downconverters), the IF is given by the following basic expression:

$$f_{IF} = f_{RF} - f_{LO} \qquad (13.1)$$

Actually, as well as this image frequency, the mixer output will include many harmonics and spurious responses. These harmonics and spurious responses can be effectively removed by filtering (notably LPF, BSF, and/or BPF, all covered in Chapter 6).

As an example, assume that the signal frequency is 38 GHz and the first IF is chosen to be 7 GHz. To what frequency must the local oscillator be tuned?

To solve this, simply rearrange (13.1) for f_{LO}, giving

$$f_{LO} = f_{RF} - f_{IF}$$

and substituting the given values:

$$f_L = 38 - 7 = 31 \text{ GHz}$$

This first local oscillator must be tuned to 31 GHz.

Following this, further downconversions lead to a final IF in the hundreds of megahertz range, typically centred on 470 MHz, and the required signal processing (usually digital, DSP) then takes place around this final center frequency.

13.2.2 Image Frequency

The image frequency is very much an unwanted by-product of the mixing (downconverting) process. It is necessary to understand what this image frequency comprises and how it becomes generated. For this purpose, first return to (13.1), rearranged to express f_{RF}:

$$f_{RF} = f_{IF} + f_{LO} \qquad (13.2)$$

Next consider an RF input signal, the frequency of which is defined as:

$$f_{image} = f_{LO} - f_{IF} \qquad (13.3)$$

(already terming this input frequency as the image frequency f_{image}).

Note the crucial change of sign here; now it is the difference between the LO and IF frequencies, whereas (13.2) for f_{RF} expresses the required sum of frequencies. Substituting the new RF input frequency into (13.1) for the IF gives

$$f_{IF} = f_{image} - f_{LO}$$

or

$$f_{IF} = f_{LO} - f_{IF} - f_{LO}$$

Giving:

$$f_{IF} = -f_{IF}$$

This is mathematically consistent because the Fourier spectrum is symmetrical about zero frequency. Although the IF appears negative, the image frequency is real and positive.

As a consequence, every receiver must be designed so as to reject this image frequency component.

A further complication is that, since f_{IF} can be negative, two possible LO frequencies are apparently available. Using (13.1) rearranged, the local oscillator frequency is given by:

$$f_{LO} = f_{RF} \pm f_{IF}$$

This fact of two possible LO frequencies clearly impacts the choice available in mixer (and general receiver) design.

The next sections deal with specific types of mixer technologies, beginning with the single-ended diode mixer.

13.3 Diode-Based Mixers

13.3.1 The Single-Ended Diode Mixer

The implementation of a single semiconductor diode (Chapter 2) is fundamental to the most basic type of mixer circuit as shown in detail in Figure 13.4.

In Figure 13.4, the RF signal v_{in} and the local oscillator source (voltage v_{LO}) are fed through the coupler and the blocking capacitor C_{B1} so that they arrive at the mixer diode, which is biased with the DC supply V_{DC}. Capacitors C_{B1} and C_{B2} are DC-blocking while inductor L_{CH} is an RF choke, minimizing any possible RF current from reaching the DC supply.

The mixing process in the diode results in a wide range of signal components and the resulting currents pass through the blocking capacitor C_{B2} to reach the LPF, which removes unwanted components, leaving only the required output, v_{IF}. It is the highly nonlinear action of the diode that is the key to the operation of this cir-

13.3 Diode-Based Mixers

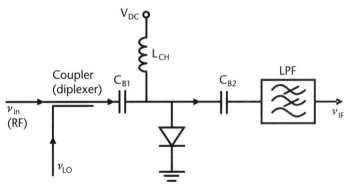

Figure 13.4 Circuit of a single-ended diode mixer.

cuit. For this reason, the next step is to analyze the behavior of this diode, activated by both the RF and local oscillator inputs.

This analysis initially follows the method given in Chapter 12 (Section 12.7), which begins by reexpressing the diode I/V expression:

$$I_F = I_{SAT}\left(e^{\alpha V} - 1\right)\} \tag{13.3}$$

which is identical to (12.9) in Chapter 12, where again all the quantities in this equation are defined in Chapter 2.

The small-signal RF (signal) and local oscillator voltages $v_{in}(t)$ and $v_{LO}(t)$ together with the resulting current $i(t)$ are indicated in the basic equivalent circuit of Figure 13.5.

Due to the nonlinear behavior of the diode, the small-signal current $i(t)$ can be expressed as a function of voltage in a series of terms in a Taylor expansion, as follows:

$$i(t) = i_0 + g_D v(t) + \left(\frac{g_D'}{2}\right)v^2(t) + \ldots \tag{13.4}$$

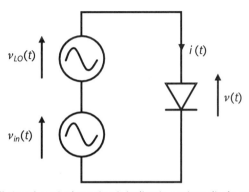

Figure 13.5 Basic small-signal equivalent circuit indicating mixer diode voltages and current.

in which i_0 is the steady-state diode current component, g_D and g'_D are, respectively, the diode transconductance and its derivative. These quantities are used to determine a closed expression for the IF component of the diode current.

In this case, it is best to use cosine waves (rather than sine waves) to express the RF and LO input voltages because this approach simplifies the trigonometric analysis. These voltages can be expressed as

$$v_{in}(t) = V_{in} \cos 2\pi f_{RF} t \tag{13.5}$$

and

$$v_{LO}(t) = V_{LO} \cos 2\pi f_{LO} t \tag{13.6}$$

It is assumed that these two waves are additive and so the next step is to substitute into the third term on the right side of (13.4), giving the IF current component:

$$i_{IF}(t) = \frac{g'_D}{2} V_{in} V_{LO} \cos 2\pi f_{IF} t \tag{13.7}$$

It can be seen that the IF current is an explicit function of the first derivative of the diode transconductance and the product of the RF and LO maximum voltage amplitudes. Also, this type of basic mixer circuit is often termed a linear mixer because (13.7) provided V_{LO} is constant then the IF current $i_{IF}(t)$ will vary linearly with the input signal voltage V_{in}.

13.3.2 The Double-Diode Mixer

The double-diode (or balanced) mixer is one step more complex than the basic single-ended diode mixer but these exhibit the following advantages: improved RF signal input matching, and improved RF-LO isolation.

For the proper operation of these types of mixers it is necessary to introduce the RF signal and LO voltages with typically 90° different phasing to produce the required mixing process and provide cancellation of the unwanted DC component. Microwave versions use 3-dB quadrature hybrids to achieve the above and a good example is the Lange coupler described in Chapter 4. In concept, the subsystem is shown in Figure 13.6.

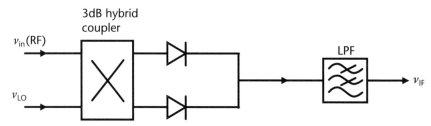

Figure 13.6 Overall subsystem of a double-diode or balanced mixer.

There are two optional configurations for the 3-dB hybrid couplers: a 90° coupler or a 180° coupler.

In the first instance (implementing a 90° coupler) the top-right hybrid output port will have 0° phase shift of the RF signal, while there will be a −90° phase shift for the signal emerging from the bottom left port.

When a 180° coupler is instead implemented, then the top-right hybrid output port will have −90° phase shift of the RF signal while there will be a +90° phase shift for the signal emerging from bottom left port. In all other respects, the overall responses are identical.

Various unwanted responses can be minimized by incorporating spur-line BSF within microstrip limbs within the hybrids (see Chapter 6).

The feature of having in-phase (0°) and quadrature (90°) versions of the RF signal or the local oscillator is synonymous with the abbreviations I and Q, used ubiquitously in many communications circuits and systems.

13.3.3 The Image-Reject Mixer

In Section 13.2.2, the issues surrounding the existence of an image frequency are dealt with in detail. In that section, it is also made clear that all receivers should be designed so that the image frequency is suppressed or (best of all) rejected. The subsystem shown in Figure 13.7 can be realized to eliminate the image frequency.

The functions of and requirements for hybrids and lowpass filters have already been described in connection with other mixer subsystems. However, in this image-reject mixer two parallel RF/IF circuits are implemented: each eventually feeding into the final output hybrid. Note that the first hybrid operates at RF while the final (output) hybrid operates at IF. It is also important to observe that the LO feeds each internal mixer circuit simultaneously. Being unused, it is important to terminate the second input port of the first hybrid (Z_0).

Each of the internal mixers is actually a complete double-diode mixer. Therefore, the entire image-reject mixer subsystem is relatively complex, comprising an overall total eight passive circuit structures (hybrids and LPFs). This arrangement is not at all compatible with IC (MMIC) manufacture.

Effectively, this circuit results in the internal isolation of the two image responses. Either of the lower sideband (LSB) and upper sideband (USB) output may be selected as the output feeding the next stage in the receiver.

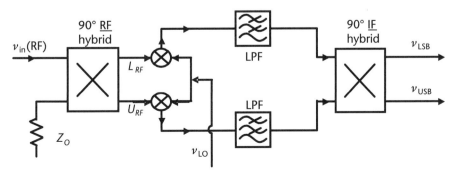

Figure 13.7 Subsystem of an image-reject mixer (LRF and URF are lower RF and upper F, channels, respectively).

An analysis of this subsystem shows that:

- With this type of image reject mixer, the conversion loss is about the same as that applying to the single-ended mixer.
- As with the 180° balanced mixer, the RF input match is good.
- The third-order intercept (IP3) is also good (i.e., low distortion: typically 20 to 30 dBm).
- Port-to-port leakage is typically around −30 dBm.
- Good isolation between RF and LO circuits.

Conversion loss is considered in Chapter 9 and the relevant section should be referred to in this context.

Diode-based mixers are often chosen for millimeter-wave applications, for example, V-band, E-band, or above where GaAs Schottky diodes are usually implemented (see Chapter 2).

13.3.4 Upconverters

In Section 13.2, the basic principle of the downconverting mixer is described. However, when the input frequencies fed to this fundamental subsystem are judiciously altered an upconverter is obtained. This is shown in Figure 13.8 (rearranged from Figure 13.3).

In effect, the RF input and the IF output, applying to a downconverter, are now interchanged. What was the IF is now the signal input and this is now mixed with the LO such that

$$f_{RF} = f_{LO} \mp f_{IF} \tag{13.8}$$

where f_{IF} contains all the signal input content, but note the sum or difference options in (13.8).

For example, assume that f_{IF} = 1.8 GHz and also that f_{LO} = 22.8 GHz is selected. What are the two possible options for the output (upconverted) signal frequency?

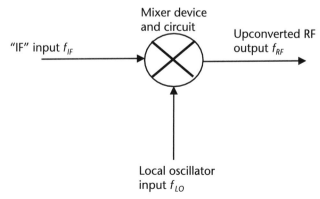

Figure 13.8 General configuration of an upconverting mixer.

Option 1: Taking the negative sign in (13.8), $f_{RF} = f_{LO} - f_{IF}$; therefore, f_{RF1} = 22.8 − 1.8 = 21 GHz.

Option 2: Taking the positive sign in (13.8), $f_{RF} = f_{LO} + f_{IF}$; therefore, f_{RF2} = 22.8 + 1.8 = 22.6 GHz.

The desired choice of frequency is then selected using appropriate BPFs.

There are several possible types of subsystem that can be considered for upconversion. One interesting option involving in-phase (I) and quadrature (Q) sources is shown in Figure 13.9.

In Figure 13.9, suitable arrangements for the frequency divider (÷2) are provided in Chapter 12. The I and Q signals are generated by digital-to-analog converters (DACs) and these are described in Chapter 11.

This subsystem reduces what is known as injection pulling, which is a potentially serious issue in conventional upconverters. Injection pulling refers to a situation in which the local oscillator frequency is changed by frequency components reflected back into the LO itself. The reduction of this highly undesirable injection pulling derives from the larger separation between LO and RF frequencies available with this arrangement. BPFs with particularly low in-band loss are desirable for use with this type of converter.

13.4 Transistor-Based Mixers

Using transistors rather than diodes opens up the possibility of signal conversion gain rather than the conversion loss associated with diode-based mixers. However, there are disadvantages, including generally inferior noise figures and somewhat degraded RF-LO isolation.

An inferior noise performance derives from the extra sources of noise associated with transistors compared with diodes (see Chapter 9). FET-based mixers use the inherently nonlinear variation of g_m as a function of V_{GS}. BJT-based mixers use the inherently nonlinear variation of g_m as a function of I_B. Specific examples of various transistor-based mixers are described in the following sections.

13.4.1 The Single-Ended FET Mixer

An example of this type of mixer circuit is shown in Figure 13.10.

In Figure 13.10, the coupler functions as described above for diode-based mixers (see Figure 13.4). Because the FET input impedance is unlikely to be anywhere

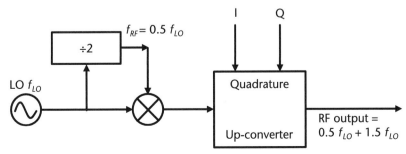

Figure 13.9 Direct frequency conversion based on halved LO frequency. (After: [4].)

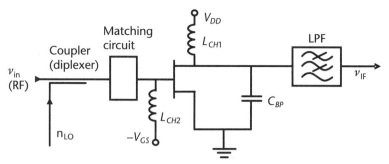

Figure 13.10 Typical configuration of a single-ended FET-based mixer (omitting FET DC biasing components).

near the impedance looking back toward the RF source, an impedance matching circuit is almost always required between the coupler and the FET input. Inductors L_{CH1} and L_{CH2} act as chokes preventing significant amounts of signal energy reaching the DC bias supplies. The shunt capacitor C_{BP} is designed to bypass any trace of the LO signal from entering the LPF.

However, transistor-based mixers implementing differential sub-circuits and baluns (see Chapter 4) are of considerable importance and these are considered next.

13.4.2 Differential FET Mixer

The general circuit arrangement for a differential FET mixer is shown in Figure 13.11.

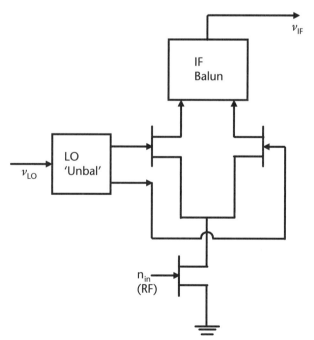

Figure 13.11 General configuration of a differential FET mixer (omitting FET DC biasing components).

13.4 Transistor-Based Mixers

The unbal is required for the single-ended LO feed. A differential LO output can feed directly, obviating the need for an unbal.

The local oscillator signal must be applied to each (differential) transistor gate, but this signal originates from an unbalanced source (i.e., the LO itself). However, the FET gates are floating (i.e., balanced); therefore, the LO signal has to be converted from unbalanced to balanced, hence the LO unbal block. Baluns are described in Chapter 4; unbals are simply baluns reversed.

Contrastingly, the RF input is already unbalanced and therefore this can be directly fed to the gate-source input of the lower FET. With the LO and RF inputs now mixed, the balanced pair of outputs (drains) from the differential pair of FETs are fed to the IF balun that converts the IF signal to an unbalanced version for feeding to the next stage which is usually the IF amplifier.

13.4.3 CMOS-Based Mixers

The design sequence for the FET mixer is as follows:

- Select the basic differential FET configuration (i.e., the core transistor configuration of Figure 13.12), ensuring that the CMOS FET process characteristics meet the design requirements, particularly that they cover at least the full ultrahigh frequency range.
- Choose suitable drain bias resistors (each being 1 kΩ in this case).
- The input impedance of the tail transistor will be anything but 50Ω, and therefore it is necessary to choose a suitable input matching circuit configuration: a simple C-L ladder network in this case.

For the design procedure, each of these components must have a unique name and also a detailed description identifying precisely its location in the circuit.

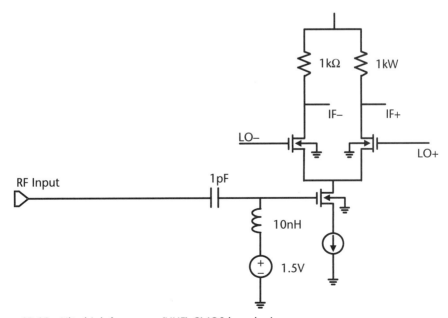

Figure 13.12 Ultrahigh frequency (UHF) CMOS-based mixer.

The completed final circuit design is indicated in Figure 13.12.

In practice the source current (the source-to-ground current generator in Figure 13.12) is periodically monitored.

This mixer operates over the 900-MHz to 950-MHz band. However, the conversion gain actually averages a loss of about 6.7 dB.

13.4.4 Mixer Implementing a Cascode Circuit

A cascode configuration of FETs can also be used as a mixer, and a basic example is shown in Figure 13.13.

In Figure 13.13, the RF and LO inputs are applied to the gates of the FETs. The RF choke and the output BPF have their usual circuit functions. The mixing action is a function of the RF and LO voltages being applied to the FET gates and experiencing the effect of the (nonlinear) transconductances of the two FETs. As a result, the final output voltage v_o is the IF.

13.4.5 The Gilbert Cell Mixer

The main basis underlying a Gilbert cell mixer is a linear circuit having time-varying parameters that enable the time-domain multiplication of two input signals. The Gilbert cell is usually realized by two differential amplifier stages each comprising emitter-coupled pairs (BJT realization) whose outputs are fed by opposite-phased signals. An example of such a circuit is provided by Figure 1.14.

Taking advantage of its symmetry and relative complexity, this circuit provides, compared with the single-ended FET mixer, the following advantages:

- Relatively high RF-LO isolation;
- Wide dynamic range;
- Cancellation of the even-order intermodulation products (IMDs);
- Cancellation of other unwanted spectral components.

Much of the following description and analysis broadly follows that provided in [4].

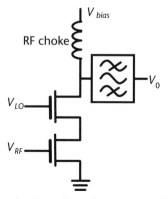

Figure 13.13 Simplified cascode circuit configured as a mixer [4]. (©Artech House, 2016.)

In Figure 1.14, the RF input signal is fed to the differential transconductance amplifier (Q_5, Q_6) the outputs of which are, in turn, fed to the common-emitter node of the internal Gilbert cell ($Q_1 - Q_4$).

Meanwhile, the LO balun outputs are fed to the (joined) bases of Q_1 and Q_4 and the bases of Q_2 and Q_3 (which are also joined).

Transistor Q_7 is permanently biased (V_{tail}) to provide the desired common DC emitter current for transistors Q_5 and Q_6.

The IF output of the Gilbert cell is the product of the two input signals but the exact situation depends on whether the circuit is configured as a downconverter or an upconverter, as follows:

- For a downconverting mixer, the IF output is the product of the LO and RF signals.
- For an upconverting mixer, the output is the product of the LO and the input signals (where the input is equivalent to the IF).

This type of double-balanced mixer exhibits conversion gain and a superior noise performance compared with other mixer configurations (particularly when SiGe HBTs are implemented). Being transistor-intensive, this circuit is also highly suited to MMIC/RFIC realization.

A simplified (equivalent) presentation of the core of the Gilbert cell is shown in Figure 13.14.

Assuming that the circuit of Figure 13.14 has an output load impedance (at the IF) Z_L, then the IF output voltage as a function of time will be:

$$v_{IF}(t) = Z_L i_{out}(t)$$

or

$$v_{IF}(t) = Z_L \alpha_F i_S \tanh\left\{\frac{v_{LO}(t)}{2V_T}\right\} \tag{13.9}$$

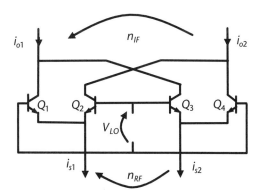

Figure 13.14 Simplified (equivalent) presentation of the core of a Gilbert cell [6]. (© Artech House, 2016.)

(A more extensive analysis is given in [7].)

In (13.9), α_F is the DC transistor collector-emitter current gain factor ($\alpha_F = I_C/I_E$ = 0.95 to 0.999); and V_T is the maximum value of v_{LO} which serves to normalize the tanh $\left\{\dfrac{v_{LO}(t)}{2V_T}\right\}$ argument and make it dimensionless.

Now i_S in (13.9) can also be written as $g_m v_{RF}(t)$ where g_m is the transconductive gain of the RF stage, so that (13.9) becomes:

$$v_{IF}(t) = Z_L \alpha_F g_m v_{RF}(t) \tanh\left\{\frac{v_{LO}(t)}{2V_T}\right\} \quad (13.10)$$

where, with currents and voltages shown in Figure 13.14, $i_S = i_{S1} - i_{S2}$ and: $i_S = g_m v_{RF}(t)$. Also,

$$i_{out} = i_{o1} - i_{o2} \quad (13.11)$$

An important (driving) aspect is that the local oscillator voltage $\{v_{LO}(t)\}$ must be high enough to drive transistors ($Q_1 - Q_4$) alternately fully into the conduction mode and completely into the off mode. This infers that $v_{LO}(t)$ must be a square wave as illustrated in Figure 13.15.

In the waveform shown in Figure 13.15, the maximum voltages could be, for example, ± 2.5 to $\pm 3V$.

With this square-wave drive, the states of the transistors ($Q_1 - Q_4$) will have negligible effects on the linearity of the mixer.

Finally and importantly, the IF output voltage can be expressed in terms of the RF voltage (maximum value V_{RFM}), the local oscillator frequency (f_{LO}), and the signal frequency (f_{RF}). This result is best developed by starting with the RF input signal as a cosine wave:

$$v_{RF}(t) = V_{RFM} \cos(2\pi f_{RF} t \quad (13.12)$$

Substituting this expression (13.12) into (13.10) and expanding the tanh function as a series of recurring cosines of sums and differences of frequencies gives the following approximate expression:

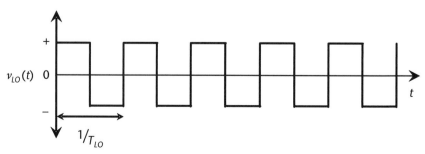

Figure 13.15 Local oscillator waveform as a square wave.

$$v_{IF}(t) = \frac{2Z_L \alpha_F g_m V_{RFM}}{\pi} \begin{Bmatrix} \cos[2\pi(f_{LO} - f_{RF})t] \\ + \cos[2\pi(f_{LO} + f_{RF})t] \end{Bmatrix} + \ldots \qquad (13.13)$$

The sums and differences occurring within this type of result are also encountered in Chapters 12 and 14.

Thus, the principal output signal comprises a component at frequency ($f_{LO} - f_{RF}$) and a component at frequency ($f_{LO} + f_{RF}$). For a downconverting mixer, the first result is required. The higher frequency component is easily filtered out using a LPF.

Originally Gilbert cell mixers were almost exclusively used as mixers in mobile (cell) phones. However, designs have been completed and operation has been proven at frequencies well into the upper E-band (i.e., at least 75 GHz).

Wideband, matched, double-balanced mixers can cover the frequency range from 3 GHz to at least 20 GHz and can either be used as downconverters or alternatively as upconverters. Designs ensure that the LO frequency can be close to that of the RF signal and LO to RF leakage specifications below −25 dBm relax the specifications required of the necessary external filters.

References

[1] Voinigescu, S., *High-Frequency Integrated Circuits*, Cambridge, U.K.: Cambridge University Press, 2013.

[2] Maas, S. A., *Nonlinear Microwave Circuits*, New York: IEEE Press, 1997.

[3] Pozar, D. M., *Microwave and RF Design of Wireless Systems*, New York: John Wiley & Sons, 2001.

[4] Camarchia, V., R. Quaglia, and M. Pirola, *Electronics for Microwave Backhaul*, Norwood, MA: Artech House, 2016.

[5] Chen, J. -D., and S. -H. Wang, "A Low-Power and High-Gain Ultra-Wideband Down-Conversion Active Mixer in 0.18-μm SiGe Bi-CMOS Technology," *IEEE Journal on Circuits, Systems and Signal Processing*, Vol. 36, No. 7, July 2017, pp. 2635–2653.

[6] Jiang, C., T. K. Johanson, and V. Krozer, "Conversion Matrix Analysis of GaAs HEMT Active Gilbert Cell Mixers," *IEEE Trans. on Microwave Theory and Techniques*, 2006, pp. 94–97.

[7] www.electronics.dit.ie/staff/ypanarin/Lecture%20Notes/DT021-4/7%20Gilbert%20Cell%20&%20Analog%20Multipliers%20(4p).pdf.

CHAPTER 14

Modulation Techniques and Technologies

14.1 Introduction

The promise of being able to communicate information using radio waves has existed for well over a century. In 1900, Landell de Moura essentially invented amplitude modulation (AM) and succeeded in transmitting (and receiving) a message using the very basic radio technology available at that time.

Eventually, AM became commercially viable and remained the de facto standard modulation technique for much of the twentieth century. Indeed, many twenty-first-century radio systems continue to use AM. In the 1920s, however, Edwin Armstrong, who had a passion for telecommunications, appreciated that varying the amplitude of a radio wave (AM) was not the only way in which to impress signal information onto the wave. Early in 1928, Armstrong started his research into the prospect of frequency modulation (FM) as an alternative to AM. The history is fascinating (and complex), but between May and October 1934, Armstrong succeeded in communicating a frequency-modulated signal between RCA's laboratory and the Empire State Building, a distance of 130 km.

By the late twentieth century, digital technology was advancing apace, and by the 1990s digital modulation techniques had entered the radio scene. Now, in the twenty-first century, digital technology has become pervasive in most areas of communications.

All modulation is based on the need for a sinusoidal carrier wave. It is this carrier wave that has to be modulated with the information desired to be communicated. Also, it is important to note that sine (or cosine) waves alone do not contain any meaningful information. This means that more complex waveforms are always involved in practice.

This chapter focuses mainly on digital modulation techniques and transceivers based on these techniques. For completeness, however, analog modulation (AM and FM) is considered first, although somewhat briefly.

14.2 Amplitude Modulation

As mentioned in the introduction, AM is the earliest form of modulating a radio wave. The basic principle is relatively straightforward: vary the amplitude of a radio wave (the carrier) proportionately with the amplitude of the signal desired to be transmitted.

A program named desmos [1] enables the user to input various wave functions, including both AM and FM, and is instructive for generating either amplitude modulated or frequency modulated waves. The user types in their preferred mathematical formula for the desired wave and the resulting waveform is displayed on-screen. For example, the waveform shown in Figure 14.1 shows the desmos output for the AM wave function of (14.1).

$$y = (1 + 0.5\cos(x+5))\sin(50x) \tag{14.1}$$

in which the variable x would contain a time element in the case of actual amplitude modulated electrical waves. The output waveform appears as shown in Figure 14.1.

A more realistic (information-bearing) AM wave is developed using the function:

$$y = (1 + 0.5\cos(x+5)0.2\cos(3x+3))\sin(50x) \tag{14.2}$$

and the resulting waveform appears as shown in Figure 14.2.

A detailed mathematical analysis of AM is now entered into.

Consider a sinusoidal carrier wave of frequency f_c and amplitude K_c given by:

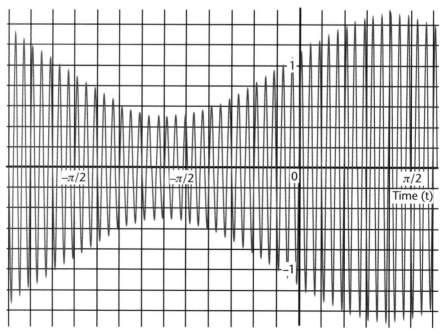

Figure 14.1 Basic AM wave generated using the desmos program.

14.2 Amplitude Modulation

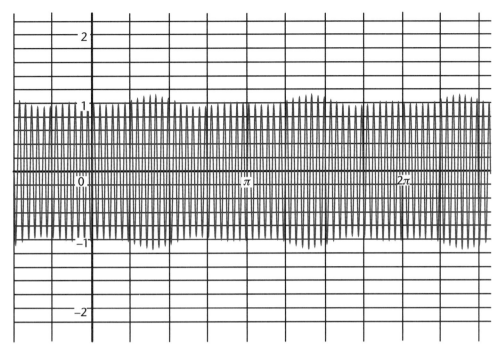

Figure 14.2 Information-bearing AM wave generated using the desmos program.

$$C(t) = K_c \sin(2\pi f_c t) \tag{14.3}$$

and further consider the modulation wave to take the function:

$$M(t) = K_m \cos(2\pi f_m t + \varphi) \tag{14.4}$$

For this initial analysis, the simplest possible modulation wave is taken (i.e., a cosine wave). In (14.4), K_m is the peak amplitude of the modulating wave, f_m is its frequency, and the phase difference φ is included because, in general, the phase will differ from that of the carrier.

It is important to observe that:

$$f_m \ll f_c \text{ always}$$

In order to preserve properly behaved modulation, K_m must always be less than 1 and $\{1+M(t)\}$ must always be positive. However, if $M(t)$ is more negative than -1, then what is known as overmodulation can occur. When AM is applied to form $P(t)$ when $\{1+M(t)\}$ multiplies the carrier $C(t)$, that is:

$$P(t) = \{1 + M(t)\} C(t)$$

Substituting from (14.3) and (14.4):

$$P(t) = \{1 + K_m \cos(2\pi f_m t + \varphi)\} K_c \sin(2\pi f_c t) \tag{14.5}$$

By expanding the main bracket in this equation and converting the trigonometric product into the addition of two sine functions (using the standard trigonometric identity) gives

$$P(t) = K_c \sin(2\pi f_c t) + \frac{K_c K_m}{2}\{\sin[2\pi(f_c + f_m)t + \varphi] + \sin[2\pi(f_c - f_m)t - \varphi]\} \quad (14.6)$$

From (14.6), it can be seen there are three spectral components:

- The carrier itself at maximum amplitude K_c and frequency f_c;
- A component at amplitude $K_c K_m/2$ and frequency $(f_c + f_m)$;
- A component at amplitude $K_c K_m/2$ and frequency $(f_c - f_m)$.

The higher-frequency component is called the upper side-frequency while the lower-frequency component is called the lower side-frequency.

In practice, modulation will cover a range of modulating signal frequencies (i.e., a band of f_m frequencies), and correspondingly there are two sidebands rather than just two side frequencies. These are therefore the upper and lower sidebands (USB and LSB) and the full spectrum is shown in Figure 14.3.

This entire signal is termed a double-sideband AM (DSB AM) signal.

A DSB AM signal can be generated by inputting the highly stable (and high-purity) carrier into one input port of a mixer while inputting the basic information signal into the other mixer input port. The output comprises the final DSB AM signal and this arrangement is shown in Figure 14.4.

Each sideband contains all the information that was within the original basic signal and therefore DSB transmission is strictly duplicating this information. If, for example, just the lower sideband (LSB) were selected, then the final transmitted signal would theoretically be complete. However, it is usually desirable for the carrier itself to be recoverable, in which case at least a vestigial version of this (i.e., low-level) is normally included. The upper sideband would be removed by filtering it out.

For example, an audio input signal having a maximum frequency f_m of 16 kHz would require at least 2 × 16 = 32 kHz of bandwidth (accommodating the carrier) for full AM transmission.

The envelope detection of an AM signal is shown in Figure 14.5.

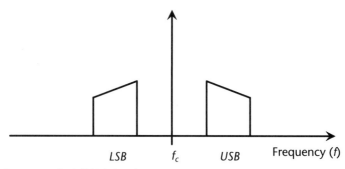

Figure 14.3 Spectrum of a full AM signal.

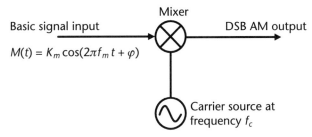

Figure 14.4 Generation of DSB AM.

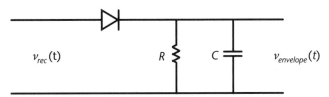

Figure 14.5 Envelope detecting circuit for an AM signal.

Due to the rectifying diode in this circuit, only the positive-going half-cycles of the carrier are detected and subsequently fed to the RC part of the circuit. The detected envelope is formed by the succession of half-cycle carrier peaks, and the result is shown in Figure 14.6.

Regarding the RC time constant (Figure 14.5), it is important for this to be:

1. Large enough so the capacitor voltage does not decrease too rapidly before the arrival of the next carrier half-cycle maximum;
2. Sufficiently small such that the output can continue to track the envelope of the signal while it decreases.

14.3 Frequency Modulation

As mentioned above, frequency modulation (FM) entered a period of rapid development in the second half of the twentieth century. By the late 1950s, FM was well

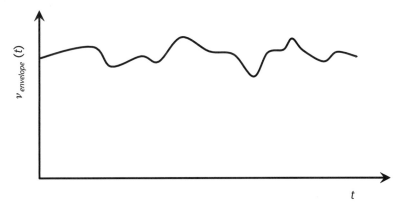

Figure 14.6 Envelope waveform detected (output from the circuit shown in Figure 14.5).

established, although the relatively wide bandwidth required meant that this form of modulation required very high frequency (VHF), 30 to 300 MHz, transmission rather than the lower frequencies associated with AM.

A carrier is first required and this can be represented by an expression very similar to (14.3), although in this instance a cosine function is used for mathematical convenience:

$$C(t) = K_c \cos(2\pi f_c t) \tag{14.7}$$

To produce FM, the carrier frequency f_c is caused to vary with time in accordance with the amplitude of the basic (information-bearing) signal input. The circuit arrangement for this, shown in Figure 14.7, depends critically on a voltage-controlled oscillator (VCO), described in Chapter 12.

The center frequency of the VCO shown in Figure 14.7 will be the value of the carrier center frequency f_c.

It is necessary to convert the frequency variations back into equivalent amplitude variations and fairly sophisticated circuits are needed to perform this function. One (commonly used) such circuit is known as the Foster-Seeley discriminator. This discriminator mainly comprises two detecting diodes, a coupling circuit and a symmetrical RC filtering network. The FM signal is fed into the coupling circuit, detected by the diodes and the resulting voltage is filtered to mainly provide the required amplitude-varying signal output.

Over a time interval t, the modulated wave $P(t)$ can be written as:

$$P(t) = K_c \cos\left\{2\pi \int_0^t f(\tau) d\tau\right\}$$

in which τ is the general time progression under the integration process. This equation can also be written as:

$$P(t) = K_c \cos[2\pi \int_0^t \{f_c + \Delta f M(\tau)\} d\tau]$$

or more compactly,

$$P(t) = K_c \cos\left[2\pi f_c t + 2\pi \Delta f \int_0^t M(\tau) d\tau\right] \tag{14.8}$$

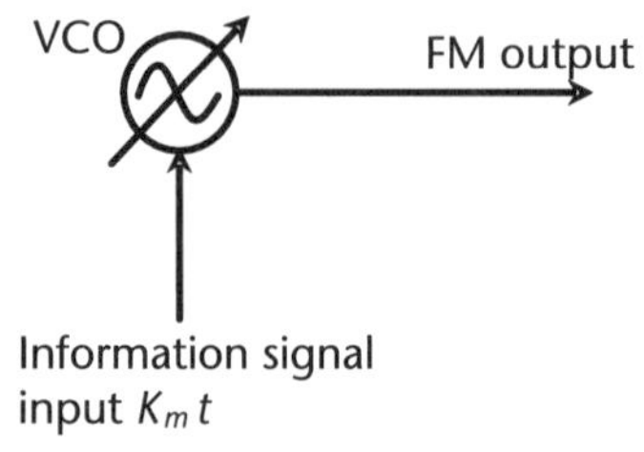

Figure 14.7 Basic arrangement for generating FM.

in which the frequency deviation $\Delta f = K_f K_m$.

K_m is defined earlier here as the peak value of the signal input and K_f is the sensitivity of the frequency modulating circuit.

A Fourier expansion of (14.8) results in a spectrum comprising an infinite number of sidebands, and this would obviously be impossible to transmit. However, in 1922, John Renshaw Carson discovered a very useful rule of thumb that largely overcomes this infinite sidebands objection. Carson found that 98% of the energy in an FM waveform is contained within a certain finite bandwidth, and the result became known as Carson's bandwidth rule:

$$B \approx 2(\Delta f + f_m) \qquad (14.9)$$

where B is the bandwidth, Δf is the peak frequency deviation, and f_m is the highest frequency component in the modulating signal.

For example, consider an FM system in which $\Delta f = 20$ kHz and f_m is 16 kHz. Determine the bandwidth required to successfully transmit and receive the FM signal. Using Carson's bandwidth rule, (14.9), the result is ≈ 72 kHz. This is approximately double the requirement for DSB AM. However, the advantage of FM is a considerably greater resistance to noise. This is because with FM the frequencies are varied sympathetically with the original information signal, while noise mainly comprises random amplitude variations. The best-known example of the commercial use of FM is almost certainly VHF FM radio.

14.4 Digital Modulation

The approach generally known as digital modulation embraces a wide range of detailed techniques and technologies, each of which has its place of importance in many modern communications systems. There is one aspect that all modulation techniques continue to have in common, namely, the need for a basic, stable, and clean (i.e., very low-noise) sinusoidal carrier wave.

Radio spectrum availability for communications is becoming increasingly limited and the pressure to conserve spectrum will only increase in future. The fact that digital modulation techniques are all considerably more spectrally efficient than analog techniques makes digital communications the first choice.

For digital modulation, the signal (a binary data stream) is caused to alter the carrier in a unique manner, depending on the exact choice of modulation. Basically, this choice amounts to three possibilities: amplitude variations, frequency variations, or phase changes, always according to whether the signal is a 1 or a 0.

14.4.1 Specific Aspects Relating to Digitally Modulated Systems

14.4.1.1 Energy per Bit and BER

Digitally modulated communications systems are fundamentally characterized by binary bit streams measured in bits/s (or bps). Since, with RF systems, relatively high-speed bit streams are usually involved the actual bit rates tend to be measured in Mbps or Gbps.

A particularly significant specification of any digital receiver is the probability of bit error, often termed the bit error rate (BER). It is important here appreciate the following general aspects:

- BER must be as low as practicable in all digital receivers (around 10^{-5} or lower).
- BER is a strong function of the energy per bit E_b and the noise power spectral density N_0. BER decreases nonlinearly as the ratio E_b/N_0 increases (exponentially in some cases and functions of the complementary error function in other instances).
- BER depends critically on the modulation scheme that is used (modulation schemes are described later in this chapter).

Pozar [2] provided an extensive treatment of bit rate and bandwidth efficiency (also termed spectral efficiency or spectrum efficiency).

The ratio E_b/N_0 is an important and fundamentally dimensionless parameter.

Since the dimensions of E_b are W-sec and those of N_0 are W/Hz, the dimensions cancel in the numerator and denominator of this ratio. E_b/N_0 is usually quoted in decibels:

$$\left(E_b/N_0\right) dB = 10 \log E_b/N_0 \tag{14.10}$$

where the bit rate of the binary signal is R_b (bps) and the RF signal power is P_s (W), the bit energy-to-noise density ratio can be expressed as

$$E_b/N_0 = P_s/N_0 R_b \tag{14.11}$$

This expression shows that the BER will always decrease as signal power increases, and/or as receiver noise decreases, and/or as bit rate decreases.

It is also useful to express E_b/N_0 in terms of the receiver signal-to-noise power ratio P_s/P_n. For a receiver having an IF bandwidth Δf and with noise having a two-sided power spectral density of $N_0/2$, the noise power is $2\Delta f\, N_0/2$, which simply equates to $\Delta f N_0$. Substituting this result into (14.11) yields

$$E_b/N_0 = \left(P_s/P_n\right)\left(2\Delta f/R_b\right) \tag{14.12}$$

The modulation scheme selected determines the bandwidth required to receive the signal. With binary modulation, one bit of information is transmitted during each bit period and such a system is defined as having a bandwidth efficiency (or spectral efficiency) of 1 bps/Hz. Higher-level modulation schemes have bandwidth efficiencies much greater than 1 bps/Hz.

14.4 Digital Modulation

Many digital communications systems adopt high-level quadrature-amplitude modulation (QAM), as high as 256QAM, 1,024QAM, or even higher. QAM is described in detail later here.

14.4.1.2 Channel Capacity

The probability of bit error (or BER) decreases strongly and nonlinearly as E_b/N_0 increases. As higher-order digital modulation schemes are introduced (e.g., 64QAM, 256QAM, or 1,024QAM), so ever-higher E_b/N_0 levels are required in order to achieve an acceptably low BER. The effects are shown for QPSK and 64QAM in Figure 14.8. The base here is the signal-to-noise power ratio P_s/P_n, which is directly related to E_b/N_0 as indicated in (14.12).

Although only two curves are shown in Figure 14.8, various other modulation schemes such as 8QAM and 16QAM exhibit characteristics between QPSK and 64QAM. Higher levels of QAM result in curves to the right of 64QAM, which means that higher signal-to-noise ratios (equivalently higher E_b/N_0 values) will be required in order to achieve sufficiently low BER. Typically acceptable BER values are well below 10^{-5}.

Because the energy per bit $E_b = P_s/R_b$ where P_s is the signal power and R_b is the data rate, there is the direct implication that for a fixed value of P_s there must be a critical value for R_b at which value the BER may become as low as desired. This critical value is termed the channel capacity (C). A very important formula providing this critical value was derived by Claude Shannon (working in conjunction with Hartley), and this formula is

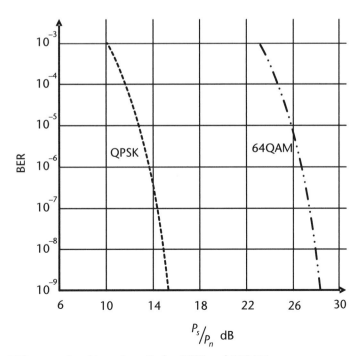

Figure 14.8 BER versus signal-to-noise ratio for QPSK and 64QAM.

$$C = B \cdot \log_2\left\{1 + \left(P_s / N_0 B\right)\right\} \quad (14.13)$$

the Shannon-Hartley theorem, where C is the maximum possible bit rate for the channel (bit/s), B is the channel bandwidth (Hz), and the remaining symbols have already been defined.

In practice, most schemes result in systems performing at only a fraction of the Shannon limit. However, the introduction of error correcting codes can cause a system to operate near to this important limit.

As described above, bandwidth efficiency is also an important consideration. With binary modulation, one bit of information is transmitted during each bit period and such a system is defined as having a bandwidth efficiency of 1 bps/Hz. Higher-level modulation schemes have bandwidth efficiencies greater than 1 bps/Hz, much higher where M-ary modulation rates of 256QAM or higher apply. For example, the LTE-Advanced standard achieves 30 bps/Hz of spectral efficiency.

Equations for BER applicable to various modulation schemes, also spectral efficiencies, are dealt with later here.

14.4.2 ASK, OOK, and FSK

Amplitude variations include amplitude-shift keying (ASK) or on-off keying (OOK), whereas frequency variations are called frequency-shift keying (FSK). Examples of waveforms associated with these three techniques are shown in Figure 14.9.

ASK and OOK are essentially extreme forms of AM and inevitably sidebands above and below the carrier are produced. The spectral extremes must include harmonics relating to the binary pulses that are modulating the carrier and the resulting overall bandwidth is therefore well over twice the individual sideband width.

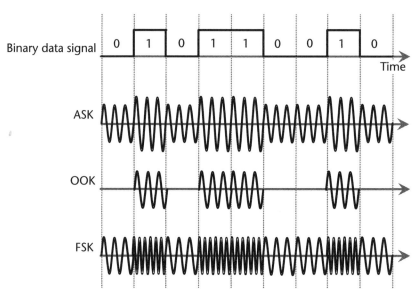

Figure 14.9 Waveforms for three basic digital modulation techniques.

Frequency-shift keying (FSK) is effectively an extreme form of FM in which the binary signal shifts the carrier frequency between relatively low and relatively high states, for the duration of the binary pulse, binary 0 or binary 1. The two resulting frequencies are referred to as the mark (1) and space (0) frequencies. As described earlier, FM generates a range of sidebands above and below the carrier frequency. The overall bandwidth depends on the highest modulating frequency, harmonics, and also the modulation index, m, given by (14.14).

$$m = \Delta f.T \qquad (14.14)$$

in which Δf is the frequency deviation between the mark and space frequencies and T is the period for each binary pulse. Δf is the difference between the space and mark frequencies:

$$\Delta f = f_s - f_m \qquad (14.15)$$

From these relations, it can be seen that as the modulation index is decreased so the number of significant sidebands is also reduced. Toward this end, minimum shift keying (MSK) is which $m = 0.5$ or as low as 0.3 is often used.

There are also two approaches that significantly increase spectral efficiency:

- Synchronize all the date signal/carrier waveform transitions.
- Decrease the harmonic content of the binary data signal.

The aim of the first of these two approaches is to ensure that all transitions (binary pulse transitions and carrier zero crossings) are synchronized in time as closely as practical.

In the second instance, the harmonic content of the binary data signal is greatly decreased through lowpass filtering. This removes the fast, sharp-edged nature of the binary signal, leaving the essential signal levels with relatively slow rise and fall times as well as smooth transitions. The example indicated in Figure 14.10 is based on the bit stream shown at the top of Figure 14.9.

Cell (mobile) phones using the GSM standard use a combination of MSK and Gaussian lowpass filtering, termed Gaussian filtered MSK (GMSK). With such phones, a data rate of 270 Kbps is accommodated within a 200-kHz bandwidth channel.

While all the above modulation techniques remain important, newer techniques in which the phase of the carrier is modulated are even more significant in modern communications systems. In the case of QAM, both amplitude and phase are used in the modulation scheme.

14.4.3 BPSK and QPSK

Binary phase-shift keying (BPSK) and quadrature phase-shift keying (QPSK) are important in their own right as well as forming a basis toward M-PSK, QAM, and other techniques. The basic waveforms relating to BPSK are shown in Figure 14.11,

Figure 14.10 Practical binary bit stream signal following LPF.

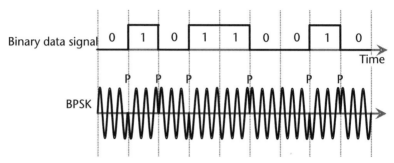

Figure 14.11 Waveforms applying to BPSK.

in which the same binary bit sequence that was used for Figure 14.9 is used again here.

Along the lower waveform in Figure 14.11, the points marked p identify the points at which the phase changes abruptly. Every transition 1 to 0 or 0 to 1 triggers a phase shift in the carrier.

The fact that with BPSK the phase transitions occur at corresponding zero-crossing points means BPSK is a coherent form of digital modulation. However, for a BPSK signal to be correctly demodulated, it requires the binary signal to be compared with a sinusoidal carrier having identical phase. This is a complex requirement, but a derivative of BPSK known as differential PSK (or DPSK) overcomes this challenge. With DPSK, the phase of each received bit is compared with the phase of the previous bit and the output from this comparison is used to determine the carrier phase change.

Basic schematic diagrams of a PSK modulator and of a synchronous PSK demodulator are shown in Figure 14.12(a, b), respectively.

For the modulation process in Figure 14.12(a), the binary signal is applied to one port of the mixer while the carrier is connected to the other input. For the demodulation process in Figure 14.12(b), the modulated signal is applied to one port of the mixer while the carrier is again connected to the other input. The mixer

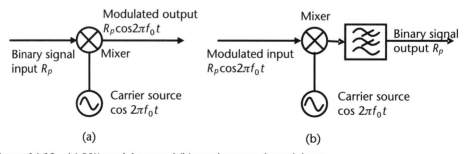

Figure 14.12 (a) PSK modulator and (b) synchronous demodulator.

14.4 Digital Modulation

output requires a lowpass filter to remove unwanted sidebands and spurious responses before the binary signal is recovered.

Technology options for mixers, carrier sources, and lowpass filters are dealt with elsewhere in this book.

QPSK takes basic PSK up to a new level. In PSK, the carrier is modulated with two phase-shift values: 0° and 180° in response to the binary 0 and 1 states. By using four phase-shift values instead of two, 2 bits can be transmitted during each signaling interval. The resulting phase states can be written as functions of the carrier as follows:

$$\text{State}_0(t) = K\cos(2\pi f_0 t + 45) \tag{14.16a}$$

$$\text{State}_1(t) = K\cos(2\pi f_0 t + 135) \tag{14.16b}$$

$$\text{State}_2(t) = K\cos(2\pi f_0 t - 135) \tag{14.16c}$$

$$\text{State}_3(t) = K\cos(2\pi f_0 t - 45) \tag{14.16d}$$

A phasor diagram exhibits all of these four states, as shown in Figure 14.13.

From this point onwards, the terms I and Q will refer to the in-phase and quadrature axes, respectively. Each phase state corresponds to a specific pairing of the binary bits in the data: 1,1 at 45°, 0,1 at 135°, and so on. Table 14.1 summarizes the complete situation.

Equations (14.16a) through (14.16d) can be comprehensively represented by the following single (general) expression:

$$\text{State}_i(t) = K_I \cos(2\pi f_0 t) + K_Q \sin(2\pi f_0 t) \tag{14.17}$$

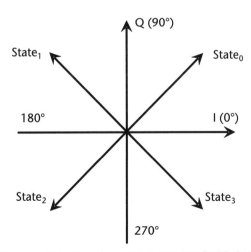

Figure 14.13 Phasor diagram of the four phase states associated with QPSK.

Table 14.1 I, Q, and Binary Data Associated with QPSK

Phase State	Phase Shift	K_I	K_Q	Bit Pairs
0	45°	1	1	1,1
1	135°	−1	1	0,1
2	−135°	−1	−1	0,0
3	−45°	1	−1	1,0

in which K_I and K_Q are, respectively, the in-phase and quadrature coefficients of the QPSK output.

This leads to the important concept of a constellation diagram for the representation of almost any type of digital modulation. For the case of QPSK, the standard constellation diagram is shown in Figure 14.14.

In practice, the phase state associations ($State_i$) are usually omitted, leaving only the bit pairs for each state.

Basically, QPSK amounts to doubling PSK, and as a result, QPSK modulators require approximately twice the number of elemental circuits compared with the PSK topologies shown in Figure 14.13. Pozar [2], for example, provided further details, but the basic principle is to take the binary data input and transfer this signal to a serial-to-parallel converter. There are two outputs from this converter, the I and Q bit streams, and each of these bit streams is fed into individual lowpass filters (LPF). The outputs from each LPF feed to the inputs of two mixers, while the I and Q versions of the carrier, $\cos(2\pi f_0 t)$ and $\sin(2\pi f_0 t)$, feed into the second inputs of each mixer. The outputs from each mixer are combined to form the final QPSK modulated output signal.

QPSK is spectrally efficient because each carrier phase is associated with two bits of data. Thus, the spectral efficiency is twice that of BPSK (i.e., 2 bits/Hz). With QPSK, twice the bit rate can be transmitted within the same bandwidth that would be required for BPSK.

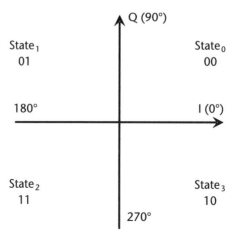

Figure 14.14 Constellation diagram of the four phase states associated with QPSK.

14.4.4 M-PSK, QAM, and APSK

14.4.4.1 M-PSK

Since it has 2 bits per state, making four amplitude-phase combinations, QPSK can alternatively be termed 4-PSK. There is no inherent reason why smaller phase shifts cannot be introduced, providing further increases in the numbers of bits per state (or symbol) that can be transmitted. For example, 16-PSK employs half-45° (i.e., 22.5°) phase shifts, each having identical carrier amplitudes. This results in 4 bits being transmitted per phase state.

Having identical carrier amplitudes is an advantage in that highly efficient but relatively nonlinear RF power amplification can be adopted.

However, as the order of magnitude of phase states increases, so the difference between the phasors decreases and the demodulator becomes increasingly subject to the effects of noise.

14.4.4.2 QAM

There is no fundamental reason why more than 2 bits cannot be transmitted with each phase state. QAM combines different carrier phases in addition to differing amplitude levels, a combination of phase shift keying and amplitude modulation.

Because of its high spectral efficiency, QAM is extensively used in cable TV networks (CATV), cellular networks (i.e., mobile phone networks), Wi-Fi LANs, and some satellite systems (SATCOM).

The lowest QAM level is 4QAM, but this is essentially identical to QPSK. The next level is called 8QAM, which embodies a total of four carrier phases together with two amplitude values. Next, there is 16QAM, which is shown in the standard constellation diagram of Figure 14.15.

With 16QAM, there are typically 12 phase shift states and 3 amplitude settings. Higher-level QAM modulation formats range as follows: 64QAM, 256QAM, 1,024QAM, and occasionally even higher. However, like M-PSK above, the demodulator becomes increasingly subject to the effects of noise, and as a result, receiver design is more complex and problematic.

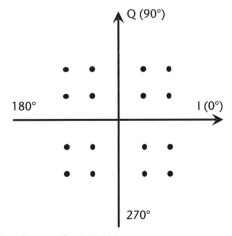

Figure 14.15 Constellation diagram for 16QAM.

14.4.4.3 APSK

There are two distinct drawbacks associated with higher-levels of QAM, namely:

- The noise disadvantage as mentioned above;
- The fact that the multitude of amplitude levels demand highly linear RF power amplifiers (RFPAs), discussed in Chapter 10, and these are always less efficient than nonlinear PAs.

Bearing in mind these drawbacks, an alternative digital modulation approach has been devised, one that avoids the multiplicity of amplitude levels, called amplitude phase shift keying (APSK). With APSK, the states are arranged in a set of concentric circles around the circumference of which the phasors are positioned, at different phases including circle-to-circle phase offsets. The resulting constellation diagram for 16APSK will exhibit fewer amplitude levels.

This arrangement of a double-ring format for 16APSK is also called 4-12 APSK because four states are located on the inner ring while 12 states are located on the outer ring.

APSK is often implemented in SATCOM systems, notably for the space segment (i.e., within spacecraft payload itself).

14.4.4.4 OFDM

Orthogonal frequency division multiplexing (OFDM) is extensively employed in many radio systems and this situation would appear to be ongoing.

This modulation scheme is designed so that although the sidebands from each carrier overlap, they can still be received without the interference that might be expected (e.g., with AM) because they are orthogonal to each another. This is achieved by having the carrier spacing equal to the reciprocal of the symbol period.

OFDM utilizes a combination of digital modulation and multiplexing in a drive to significantly improve spectral efficiency. The approach is as follows:

- Divide a predesignated transmission channel into many groups of subcarriers or subchannels.
- Choose the subcarrier frequencies and spacings so they are orthogonal to each other.

Using this approach ensures the avoidance of guard bands because the groups of spectra will not interfere with each other. The IEEE Standard 802.11n for Wi-Fi is an important OFDM case in point. The details are beyond the scope of this book, but [3] covered OFDM well.

In this standard, the subcarrier spacings and bandwidths are identical, being 312.5 kHz. Each subcarrier can be modulated using any of the digital modulation schemes described in the foregoing sections here. When 64QAM is selected, data rates as high as 300 Mbps are feasible and with higher level QAM Gbps data rates can be achieved. To obtain these advantages, fast Fourier transforms (FFT) and inverse fast Fourier transforms (IFFT) are required, as indicated in Figure 14.16.

14.4 Digital Modulation

Figure 14.16 Application of IFFT and FFT to an OFDM-modulated and multiplexed system.

The combination of a wide bandwidth and a substantial number of subcarriers results in OFDM being more resistant to signal fading (e.g., Rayleigh fading) than the other digital modulation methods used alone.

14.4.5 Spectral Efficiency of the Various Digital Systems

The very important concept of spectral efficiency is discussed in Section 14.4.1. Table 14.2 summarizes the maximum possible values of spectral efficiencies that can be obtained using the various modulation techniques.

14.4.6 Probability of Bit Error or Bit Error Rates

The probability of a received bit error (p_e) or the BER is considered in this section. The term BER is most commonly encountered in practical systems design and measurement scenarios. First, the equations for this important quantity are provided, relating to most of the modulation schemes described above. In all cases, the formula relates to the probability of error when a zero (0) is transmitted but a (1) is received, the terms are defined in Section 14.4.1 and the function $erfc(x)$ is the complementary error function (approximate closed-form expressions for this are provided in Section 14.4.7). Equations for BER are available in most standard texts, and the results for five modulation schemes are presented next.

14.4.6.1 BER for ASK Modulation

Equation (14.18) is the probability of bit error for this type of modulation.

Table 14.2 Spectral Efficiencies for Various Digital Modulation Techniques

Modulation Type	Spectral Efficiency (bps/Hz)	Comments
FSK	<1	Depends on the modulation index
GMSK	1.35	
BPSK	1	
QPSK	2	
16QAM	4	
64QAM	6	
256QAM	8	
OFDM	>10	Can be as high as 30 bps/Hz

$$P_e = 0.5 erfc\left(\sqrt{\frac{E_b}{4n_0}}\right) \qquad (14.18)$$

14.4.6.2 BER for Synchronous PSK Modulation

Equation (14.19) is the probability of bit error for this type of modulation.

$$P_e = 0.5 erfc\left(\sqrt{\frac{E_b}{n_0}}\right) \qquad (14.19)$$

Note that the argument under the root is exactly four times that applying to the ASK modulation case.

14.4.6.3 BER for Synchronous FSK Modulation

Equation (14.20) is the probability of bit error for this type of modulation.

$$P_e = 0.5 erfc\left(\sqrt{\frac{E_b}{2n_0}}\right) \qquad (14.20)$$

Note that in this FSK case the argument under the root is exactly twice that applying to the ASK modulation case.

14.4.6.4 BER for QPSK Modulation

Equation (14.21) is the probability of bit error for this type of modulation.

$$P_e = 0.5 erfc\left(\sqrt{\frac{E_s}{2n_0}}\right) \qquad (14.21)$$

In which the quantity E_s is the energy per symbol.

14.4.6.5 BER for 16QAM Modulation

Equation (14.22) is the probability of bit error for this type of modulation.

$$p_e = 1.5 erfc\left(\sqrt{\frac{2E_b}{5n_0}}\right) \qquad (14.22)$$

in which the quantity E_b is the energy per bit.
Note that this expression contrasts with the earlier formulas as follows:

- The coefficient (1.5) is three times that of the earlier formulas.

14.4 Digital Modulation

- The energy per bit is doubled under the root argument.
- The noise density n_0 is quintupled under the root argument.

These features must be taken into account for all calculations relating to BER for QAM modulation schemes.

14.4.7 Closed-Form Expressions for the Complementary Error Function

$$erfc(x) \approx 1 - \frac{3.372x}{3+x^2} \text{ where } x \leq 0.96$$

$$erfc(x) \approx \frac{1.132xe^{-x^2}}{1+2x^2} \text{ where } x \leq 2.71$$

For particular values of the argument ($x = 0, 1, 2,$ and 3), the following results apply:

$$erfc(0) = 1 \qquad erfc(1) = 1.573 \times 10^{-1}$$
$$erfc(2) = 4.678 \times 10^{-3} \quad erfc(3) = 2.209 \times 10^{-5}$$
$$\text{also } erfc(\infty) = 0$$

14.4.8 BER Data Compared

The BER values obtained using various digital modulation schemes must be compared. This comparison will always involve the carrier-to-noise ratio (dB) and how this relates to the bit energy-to-noise E_b/N_0 ratio. The required relationship is given as (14.23).

$$C/N = \left(\frac{E_b}{N_0}\right)\left(\frac{R_B}{B}\right) \tag{14.23}$$

where R_B is the channel data rate and B is the bandwidth.

In decibels:

$$C/N \text{ dB} = 10\log_{10}\left(\frac{E_b}{N_0}\right) + 10\log_{10}\left(\frac{R_B}{B}\right) \text{ dB} \tag{14.24}$$

The overall result is that, for a desired BER, a higher C/N is required as the modulation level increases (BPSK, QPSK, and through the progression of QAM levels). This infers ever-increasing carrier power requirements, resulting in RFPAs capable of delivering these increased RF power levels (Chapter 10).

It is important to appreciate that all the results shown above for spectral efficiency and BER values (Section 14.4.5 through this section) will be degraded in the presence of Rayleigh fading and additive Gaussian white noise (AGWN).

Careful antenna positioning can substantially decrease the onset of serious multipath propagation and as in all communications systems noise must always be minimized (noise is considered in Chapter 9).

A comparison of BER characteristics for QPSK and 64QAM is given as Figure 14.17.

14.4.9 Spread-Spectrum Modulation

Spread-spectrum modulation (also known as code-division multiple access [CDMA]) is a particularly important form of digital modulation. The term spread spectrum arises from the fact that with this technique the digital signal is spread over a relatively wide bandwidth but at a lower level of RF power. Another more complete name for this approach is direct sequence spread spectrum (DSSS). A block diagram showing the principle of spread-spectrum modulation is provided in Figure 14.18.

The digital input signal input (A) comprises the digitized and compressed (analog) voice waveform and this is fed to one input of the exclusive-OR (XOR) gate. What is known as a chipping signal is fed to the second gate input. The data rate of this chipping signal is two orders of magnitude higher than that of the digital signal input.

Numerical details of this system follow what is known as the CDMA IS-95 standard, which uses a chipping signal at 1.2288 Mbps and a digitized and compressed (originally voice) signal running at 13 Kbps. What is called a code generator in Figure 14.18 is actually a pseudorandom code generator that assigns a unique code to each user of the channel.

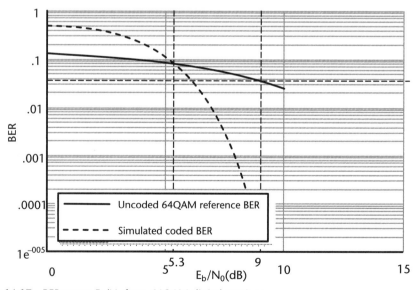

Figure 14.17 BER versus E_b/N_0 for a 64QAM digital receiver.

14.4 Digital Modulation

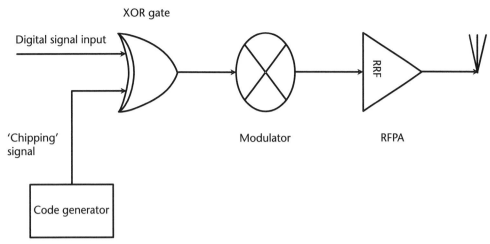

Figure 14.18 Basic circuit arrangements comprising a spread spectrum system.

Recollecting the behavior of the exclusive-OR gate (Chapter 5) and then examining representative waveforms assists in understanding how the spread-spectrum system operates.

First, the output (Q) of the exclusive-OR gate will only trigger to a binary 1 when either input is also a binary 1. All other conditions are excluded. Because pulses from the digitized signal input (S) are much longer than those associated with the pseudorandom code generator (C), there will always be a substantial number of opportunities for either S or C to equal 1 and hence a gate output Q will be triggered.

The pulse length for the 13 Kbps digitized signal input (S) is 77 μs. In contrast, the pulse length for the 1.22883 Mbps chipping signal input (C) is 0.814 μs. Synchronized waveforms representing the digitized signal input and the chipping signal are illustrated in Figure 14.19.

Clearly, at maximum, there could be 77/0.814 = 94.62 chipping pulses coincident with each 77 μs digitized signal input pulse. However, in practice, there will be randomized (coded) numbers of chipping pulses coincident with each digitized signal input pulse.

In Figure 14.19, the states between vertical lines labeled with dash-dots are different (i.e., a 1 and a 0). For this 0.814-μs pulse duration, at least the XOR gate will deliver an output 1. Contrastingly, the states between vertical lines labeled dots are identical (both 1 in the case identified), and therefore the XOR gate will deliver an output 0.

This spread-spectrum modulator spreads the original voice signal over a bandwidth of 1.25 MHz and the final low-power signal at the output looks randomized (i.e., very like noise). It is therefore extremely difficult if not impossible for an unwanted listener to uncover the code and hence decode the original message. Many users can occupy the same 1.25-MHz bandwidth simultaneously because each user simply needs his or her original code to decipher the message. A correlator at the receiver identifies and recovers a specific user's code.

3G (mobile) cellular networks use a spread-spectrum technique known as wideband CDMA (WCDMA). With WCDMA, the analog voice signal is compressed in

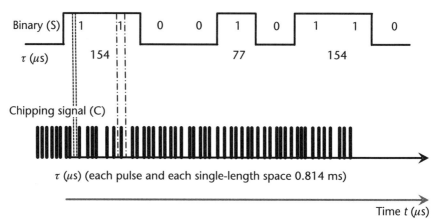

Figure 14.19 Digitized signal input and binary (S) and chipping signal (C): synchronized waveforms for spread-spectrum modulation.

much the same way as described above and a 3.84-Mbps chipping code is applied. As a result, a 5-MHz-wide channel enables many users to share this same bandwidth. China, for example, has taken up this format.

14.4.10 Orthogonal Frequency Division Multiple Access

OFDM is described above. In that section, the IEEE Standard 802.11n is referred to, in which 56 subcarriers are interleaved to occupy an overall bandwidth of 20 MHz. With OFDM, the subcarrier frequencies and spacings are chosen so they are orthogonal to each other, which avoids the need for guard bands because the groups of spectra will not interfere with each other.

For orthogonal frequency division multiple access (OFDMA), the same basic arrangement is used and it is possible to implement up to several thousand subcarriers this way, each being 15 kHz wide. The LTE (long-term evolution) cellular network standard is a good example of the OFDMA approach. The details of these principles are beyond the scope of this book, but they are covered well in [3].

14.5 Transceivers

14.5.1 Basic Concept of a Transceiver

An RF transceiver can be regarded as a complete system but the transmitter and receiver sections are usually treated as separate entities within the system as illustrated in Figure 14.20.

The signal to be communicated enters the transmitter part and is subsequently processed until a replica of this signal is power-amplified and fed to the transmitting antenna. In the receiver part the signal is again processed until the version required is finally output and fed to the receive-end computer, cell phone, sound speaker, or other final output device.

For relatively low-power systems operating at frequencies up to at least 6 GHz, complete transceivers are implemented in silicon CMOS ICs.

14.5 Transceivers

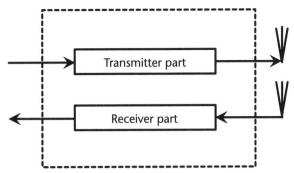

Figure 14.20 Basic concept of a combined transmitter-receiver.

14.5.2 Software-Defined Radio

This concept is briefly introduced in Chapter 1.

14.5.3 Full-Duplex Radios

Multipath and related signal degradation represents serious challenges in all radio system design. Spectrum efficiency is also a challenge, and as a result, much effort continues to be placed on solutions to these important issues.

One promising solution has been published by Yang et al. [5], who described RF and digital self-interference cancellation for full-duplex radios.

The main transceiver, central to this system, is more or less conventional and it implements circuitry described in Chapters 2 through 13.

In this system, direct RF correlative detection is employed. With this technique, the automatic adjustment of phase and magnitude optimizes the cancellation process. Correlative detection provides both the magnitude and phase information related to self-interference such that an adaptive adjustment algorithm can be readily implemented. One example of such an algorithm is the LMS algorithm.

Also, the RF cancellation module can readily be cascaded in support of multiple-input, multiple-output (MIMO) systems.

From these data, it can clearly be seen that isolation better than 50 dB is achieved over a 50-MHz bandwidth. For frequencies up to at least 6 GHz, these full-duplex transceivers are implemented in silicon CMOS ICs. The full article by Yang et al. [5] should be consulted for further details.

14.5.4 Transceiver Modules for Short-Range Radio

The Internet of Things (IoT) is providing a powerful driving force for transceiver modules applying to short-range radio links. Examples include Bluetooth, LoRa, and ZigBee. Bluetooth is now very well known, notably for its in-vehicle connection capability. ZigBee is also becoming popular, although LoRa is less familiar to many.

Most of the circuit elements involved here are described in detail in several chapters of this book. The blocks labeled S/P or P/S are phase separators that output in-phase (I) and in-quadrature (Q) signals, half-sine shaping blocks are essentially half-wave rectifiers and the summation block (Σ) combines the final I and Q signals to form the transmitter output. Following this, an RF power amplifier is usually required (although the power level will usually only amount to some hundreds of milliwatts in this application).

References

[1] https://www.desmos.com/calculator/nqfu5lxaij.

[2] Pozar, D. M., *Microwave and RF Design of Wireless Systems*, New York: John Wiley & Sons, 2001.

[3] Electronic Design Library: Focus on Wireless Fundamentals for Electronic Engineers. ED_WirelessFundamentalsEbook_May 2017.pdf.

[4] NI AWR example, http://kb.awr.com/display/Examples/DigitalAM_BER#gsc.tab=0.

[5] Yang, B., et al., "Low Complexity, High Performance RF Self-Interference Cancellation for Full-Duplex Radios," *Microwave Journal*, April 2017, pp 86–96.

Appendix A
Logarithmic Units

Anyone observing a description of almost anything to do with radio systems will rapidly appreciate that many parameters are defined using units such as dB, dBm, dBc, and occasionally dBW. All these types of units are logarithmic. Yet, given that all the defining quantities associated with RF systems could in theory at least be expressed as linear factors the question naturally arises: Why use logarithmic units or why not simply stay with linear quantities? In this appendix, the important logarithmic units (and their handling) are described in detail.

First, the reasons for not generally staying with linear ratios are:

- Power levels often differ by several orders of magnitude.
- Staying with linear units leads to often very difficult calculations.
- Using logarithmic units (dB, dBm) means direct additions and subtractions can be used all through even a complex system.

In particular, power ratios, power levels, noise levels on so on are all usually calculated, quoted and measured in dB or dBm. It all really began with the concept of the Bell, which itself goes right back to the mid-nineteenth century. In 1876, at the age of just 29, Alexander Graham Bell invented an interesting device that he called the telephone. The Bell logarithmic unit was named after Alexander Graham Bell.

A power ratio (linearly P_1/P_2) converted to x in Bells (units B) is expressed as:

$$x = \log_{10} P_1/P_2 \text{ B} \qquad (A.1)$$

From this point onward the base (10) will not be stated but instead will be assumed because all logarithmic units used in this context are common logarithms (i.e., base 10).

However, this basic ratio is cumbersome and inconvenient in practice so 10 times the log is used to form decibels (dB) and any value dB is therefore:

$$y = 10\log_{10} P_1/P_2 \text{ dB} \quad (A.2)$$

This expression is uniformly important and all RF engineers need to be immediately familiar with this formula together with some specific results.

Example A.1

Calculate the power ratio in dB where the linear power ratio is 400.

Using (A.2), the result is 26.021 dB.

However, the decimal part of this result is superfluous and in this case three significant figures are adequate, so the result may be stated as 26.0 dB with sufficient accuracy.

Example A.2

Calculate the output/input power ratio in decibels for an attenuator in which the linear power ratio is 0.3.

Again use (A.2), substituting

$$y = 10\log_{10} 0.3$$

Hence, $y = -5.23$ dB

Linear reductions in power result in negative decibel values.

Logarithmic units related to dB are also employed for defining absolute power levels (pW, nW, μW, mW, W). The most important unit of power in many electronic systems (certainly including RF) is the milliwatt (mW), and 1 mW of power is very often used as the reference level for defining power levels in dBm (decibels referenced to 1 mW). Referring to (A.2), the power in the denominator (P_2) is now fixed at this 1-mW reference level, giving (since 1 mW = 1×10^{-3} W):

$$p = 10\log_{10} P_1/(1 \times 10^{-3}) \text{ dBm} \quad (A.3)$$

which, when expanded out, becomes:

$$p = 10\log_{10} P_1 + 30 \text{ dBm} \quad (A.4)$$

where P_1 is the power level in watts.

Example A.3

Calculate the power level in dBm when the linear power level is 15 mW.

Using (A.4), the first term becomes $10\log_{10}(15 \times 10^{-3})$, which leads to the result:

$$p = 11.76 - 30 + 30$$

that is, $p = 11.8$ dBm (truncating to three significant figures)
This type of result is occasionally expressed verbally as: 11.8 dB above 1 mW.

Example A.4

Calculate the power level in dBm when the linear power level is $7\,\mu$W.

Again use (A.4), but the first term now becomes $10\log_{10}(7 \times 10^{-6})$, which leads to the result:

$$p = 8.45 - 60 + 30$$

that is, $p = -21.6$ dBm.
Power levels below 1 mW result in negative dBm values.

For relatively high-power systems, dBW units are sometimes used rather than dBm. Now the reference power level is 1W in place of the 1 mW for dBm.

Summarizing, dB, dBm, and occasionally dBW are extensively employed in radio communications and in radar systems design and assessment. These units are also used in the related specifications and instrumentation as well as in supporting EDA software.

Appendix B
S-Parameters and X-Parameters

B.1 Scattering (S)-Parameters

S-parameters are used extensively in microwave devices and systems. The basis of S-parameters (for a two-port system) starts by considering root-power waves and representing signals entering and leaving the ports. This is shown in Figure B.1.

It is important to appreciate that the device is treated as behaving approximately linearly here, even though many devices (certainly all types of transistors) behave very nonlinearly in reality. However, when operating under small-signal conditions around a fixed DC bias point, transistors may be considered to be behaving in an approximately linear manner.

To understand the concept of root-power waves (a and b), recollect that it is always the signal power levels that are of greatest importance in RF/microwave systems, but also bearing in mind that effectively the equivalent voltages will be needed at each port. Since voltage is proportional to $\sqrt{power}$, this leads to the root-power-wave concept.

In this appendix only two-port S-parameters are considered (the concept can be extended to an N-port network). Pozar [1] has provided further extensive details.

The S-parameters express the reflected root-power waves b_1 and b_2 in terms of the incident root-power waves   and   and in each instance the terms are additive. This leads to the following pair of linear equations:

$$b_1 = S_{11}a_1 + S_{12}a_2 \tag{B.1}$$

$$b_2 = S_{21}a_1 + S_{22}a_2 \tag{B.2}$$

which can be conveniently expressed by the following matrix:

$$\mathbf{b} = \mathbf{Sa} \tag{B.3}$$

The physical meanings of each S-parameter are as follows:

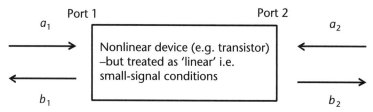

Figure B.1 Root-power waves at the input (1) and output (2) ports of a two-port linear system.

- S_{11} is the input reflection coefficient (equal to Γ_1), with the output matched.
- S_{12} is the reverse transfer parameter, with the input matched.
- S_{21} is the forward transfer parameter, with the output matched.
- S_{22} is the output reflection coefficient (equal to Γ_2), with the input matched.

The square of the forward transfer parameter (notably $|S_{21}^2|$) relates strongly to power transfer through the network or system, for example, insertion loss through a filter or the power gain of an amplifier.

The reverse transfer parameter S_{12} is very important in amplifier stability considerations.

All four S-parameters are required for detailed determinations of power gain in amplifiers, detailed examinations of amplifier stability, and computations for amplifier design.

To determine each S-parameter, it is vital to correctly terminate the input and output ports (i.e., ensure each port is matched with a Z_0 load, almost always equal to 50Ω).

Each parameter varies with frequency, DC bias point, device-to-device (even across a batch of devices), and temperature.

Many engineers are familiar with other types of two-port parameters and these are compared with S-parameters in Figure B.2.

At lower frequencies (audio through very high frequency) hybrid [h], impedance [z] or chain [ABCD] matrices are often encountered. Through ultrahigh frequency (i.e., into the microwave bands) admittance [y] parameters are sometimes used, but the required accurate short circuits are generally difficult to realize and short circuits can lead to instability during measurements. S-parameters are more robust and more readily measured.

B.2 X-Parameters

Although the small-signal S-parameters are extensively used for characterizing many active devices and circuits, these parameters do not accommodate nonlinear effects and are therefore unsuitable for use in, for example, power amplifier design.

It has therefore long been appreciated that a more comprehensive set of parameters, capable of handling nonlinearities, would be a very useful tool. Research conducted in the 1990s formed the foundation to the development of useful nonlinear parameters, but in 2006 what was then Agilent Technologies (now Keysight Technologies) made substantial strides by introducing what were coined X-parameters.

B.2 X-Parameters

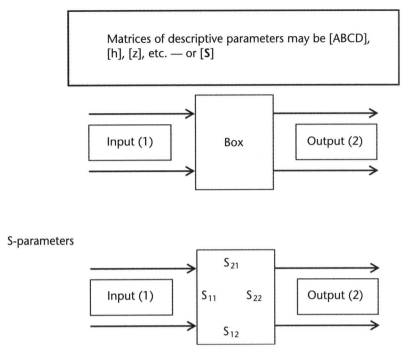

Figure B.2 Two-port (box) parameters compared.

These parameters take into account the harmonic components that are always associated with nonlinear characteristics and early work used terminology such as polyharmonic distortion. A comprehensive account of these developments is provided by David Vye [2], which includes details of the X-parameter approach. Further, highly detailed, information is contained in the MicroApps PDF by D. E. Root [3].

X-parameters are a super set of S-parameters, and the general concept is shown in Figure B.3.

In this system, nonlinear effects are accommodated and the quantities A_{pk} and B_{pk} account for this. In each case the subscript p refers to the numbered port and the subscript k refers to the kth harmonic. In this way each quantity is designed to take care of every significant harmonic and appears as a term in an overall equation. There may well be a substantial number of harmonics with significant amplitudes to consider.

The overall result comprises a set of relatively complicated equations, for example, expressing $B_{e,f}$ [3]:

$$B_{e,f} = \underbrace{X_{ef}^{(F)}(|A_{11}|)P^f}_{\text{Perfect match}} + \underbrace{\Sigma_{g,h} X_{ef,gh}^{(S)}(|A_{11}|)P^{f-h}.A_{gh}}_{\text{Mismatch}} + \underbrace{\Sigma_{g,h} X_{ef,gh}^{(T)}(|A_{11}|)P^{f+h}.A_{gh}^*}_{\text{Additional mismatch terms}} \quad (B.4)$$

where the P function accounts for the relative phases, for example, for the phase (φA_{11}) of the A_{11} wave:

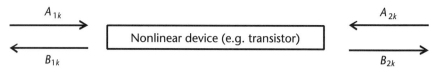

Figure B.3 Root-power waves at the input (1) and output (2) ports of a two-port nonlinear system.

$$P = e^{j\varphi A_{11}} \qquad (B.5)$$

A_{gh}^* is the complex conjugate of A_{gh}.

Root [3] has provided further details and in particular indicates:

- The simplest X-parameters for an amplifier;
- X-parameters with arbitrary load-dependence;
- Several practical examples.

In many instances of microwave and millimeter-wave design and measurement S-parameters will likely continue to be applied. Where power amplifiers are under consideration (or other applications in which nonlinearities must be accounted for), X-parameters will increasingly be utilized.

Most EDA software suites provide X-parameter callout options, including Keysight Technologies' ADS and NI AWR.

References

[1] Pozar, D. M., *Microwave and RF Design of Wireless Systems*, New York: John Wiley & Sons, 2001.

[2] Vye, D., "Fundamentally Changing Nonlinear Microwave Design," *Microwave Journal*, March 2010, pp. 22–44.

[3] Root, D. E., X-parametersMicroAppsRootFinal.pdf, May, 26, 2010.

Acronyms and Abbreviations

2-DEG	two-dimensional gas
4G	fourth generation mobile (cell phone) systems
A/D (ADC)	analog-to-digital converter
ACPR	adjacent channel power ratio
ADSL	asymmetric digital subscriber line
AESA	active electronically scanned array (increasingly used for many modern radars)
AM	amplitude modulation
APD	analog predistortion
ASIC	application-specific integrated circuit
AWGN	additive white Gaussian noise
BER	bit error rate
BiCMOS	combination of bipolar transistors and CMOS
Bipolar	bipolar transistor
BJT	bipolar junction transistor
CATV	cable television
CDMA	code-division multiple access
CIMR	carrier-to-intermodulation ratio
CMOS	complementary metal oxide silicon
COTS	commercial off the shelf
CW	continuous-wave
DAC	digital-to-analog converter
DPD	digital predistortion
DSP	digital signal processor
ENOB	equivalent number of bits
EVM	error vector magnitude
FDM	frequency division multiplexing
FET	field-effect transistor
FM	frequency modulation
FoM	figure of merit
FPGA	field-programmable gate array
GaAs	gallium arsenide, still the most commonly encountered CS semiconductor
GaAs FET	gallium arsenide field-effect transistor
GaAs MESFET	gallium arsenide metal-semiconductor field-effect transistor

GaN	gallium nitride, an increasingly important wide bandgap CS
GaN FET	a MESFET based upon GaN
GaN HBT	a heterojunction bipolar transistor based upon GaN
GaN HEMT	a HEMT based upon GaN
GSM	global system for mobile communications
HBT	heterojunction bipolar transistor
HEMT	high electron mobility transistor
HMIC	hybrid microwave integrated circuit, same as MIC
HPA	high power amplifier (RF in this context)
IF	intermediate frequency
IIP3	third-order input intercept point
IMD	intermodulation distortion
InP	indium phosphide
LAN	local area network
LDMOS	laterally diffused metal-oxide-silicon (power transistor)
LNA	low-noise amplifier, generally found in receivers
LO	local oscillator
LPF	lowpass filter (or filtering)
LTE	long term evolution (of cellular network standards)
MEMS	microelectromechanical systems (ICs which embody moving parts)
MESFET	metal-semiconductor field-effect transistor
MIC	microwave integrated circuit (usually means hybrid technology), same as HMIC
MIM	metal insulator metal
MIMO	multiple-input multiple-output
MMIC	monolithic microwave integrated circuit
MMW	millimeter-wave
MOSFET	metal-oxide-silicon field effect transistor
MUX	multiplexer
MW	microwave
MWS	microwave wideband system
NCO	numerically controlled oscillator
NF	noise figure (usually applying to a receiver or an LNA)
OFDM	orthogonal frequency-division multiplexing
OIP3	third-order output intercept point
PA	power amplifier (RF in this context)
PAE	power-added efficiency (important with RF power amplifiers)
PAPR	peak to average power ratio
PCB	printed circuit board
PDF	portable data file; probability density function
pHEMT	pseudomorphic HEMT
PIN	p-type intrinsic n-type, a passive RF diode capable of several circuit functions
PLL	phase-locked loop
PLO	phase-locked oscillator

QAM	quadrature amplitude modulation
QPSK	quadrature phase-shift keying
RF	radio frequency (any frequency above about 20 kHz, up to several hundred gigahertz)
RFIC	radio-frequency integrated circuit (typically above about 400 MHz and up to around 4 GHz); for separation purposes, we are identifying Si RFICs, but many in the industry do not distinguish between MMICs and RFICs
RFID	radio frequency identification
RFPA	RF power amplifier
RMS	root mean square
RX	receiver
SATCOM	satellite communications
SAW	surface acoustic wave
SC	single carrier
SDR	software-defined radio
Si	silicon
SiC	silicon carbide (often used as a substrate for GaN devices)
SiGe	silicon-germanium (an increasingly significant compound semiconductor)
SiP	silicon integrated package
SINAD	signal, noise and distortion
SNR	signal-to-noise ratio
SiP	silicon integrated package
SOC	system on a chip
SoP	system-on-package
SOT	small outline transistor
sps	samples per second (used in digital signal processing, hence Msps and Gsps)
SSB	single sideband
SSPA	solid-state power amplifier (RF in this context)
TDM	time-division multiplexing
TEM	transverse electromagnetic
TR	transceiver
TRM	transmit-receive module
TX	transmitter
TxRx	transmit-receive (module = transceiver, RF in this context)
UWB	ultrawideband
Varactor	variable reactor, a semiconductor diode producing a voltage-variable capacitance
VCO	voltage-controlled oscillator
WBG	wide bandgap semiconductor (e.g., GaN, SiC)
WCDMA	wideband code-division multiple access
WiMAX	worldwide interoperability for microwave access

About the Author

Terry Edwards is very well known globally for his RF/microwave expertise in the academic sphere as well as across the industry. For him, it all began when (germanium) transistors always had three wires sticking out of an enclosing aluminum can and stopped working at frequencies well below 1 MHz. Except for detectors, the microwave world meant specialized vacuum tubes and metal waveguides. What a tremendous contrast with today's GaN MMICs into, for example, multimodule digitally modulated communications systems.

Holding an M. Phil. for microwave research, his experience is highly varied, ranging from systems development with Ultra Electronics (United Kingdom) to university lecturing and research before taking on industrial consultancy, focusing largely on the worldwide RF/microwave industry. Mr. Edwards lectured at the postgraduate level in Australia for 2 years and has delivered many presentations on his main subject in the United States and the United Kingdom. He is probably best known for his book *Foundations for Microstrip Circuit Design*, of which the third and fourth editions were written in partnership with Professor Michael Steer of NCSU.

Mr. Edwards' series of lectures for the radio frequency systems M.Sc. course at Hull University (United Kingdom) provided the springboard for this current book.

Index

4G, 273
5G, 5, 7, 118, 138, 151, 182, 187

A

Active electronically scanned arrays (AESAs), 117, 119-121, 273
Active mixers, 15-16, 226, 230-239
Additive Gaussian white noise (AGWN), 260
Adjacent channel power ratio (ACPR), 199-200, 207, 273
Admittance, 131, 134, 157, 270
Air bridge, 20, 46, 67, 71, 73, 136
Aliasing, 191, 197-199
Amplifier
    Automatic gain control (AGC), 124, 129-131, 225
    Balanced, 135-136, 181
    Basic, 123, 124, 125-127
    Broadband, 134-139
    Class A, 165, 168, 170
    Class AB, 170-171
    Class B, 170-171
    Class C, 171
    Class E, 171-175
    Class F, 175-178
    Distortion, 183-186
    Distributed, 134-139
    Doherty, 178-180
    Envelope-tracking, 180-181
    Low-noise (LNA), 120, 124, 141-160
    Phase noise, 10, 41, 149-150
    Power-added efficiency (PAE), 164-166
    Power gain, 123-124, 126, 129-130, 132, 137, 138, 143, 151, 152, 153, 154, 161-166, 170-171, 270
    Push-pull, 73, 170, 181
    Stability, 127-129, 138, 151

Amplitude modulation (AM), 150, 241-245
Analog predistortion (APD), 273
Analog-to-digital converters (ADCs)
    Aliasing, 197-200
    Architectures, 189, 200-203
    Baseband analog sampling, 190
    Effective number of bits (ENOB), 191-193
    Flash, 200-201
    Folding, 201
    Pipelined, 201-202
    Quantization, 14, 189-190, 192-207
    Quantization noise, 193
    Sample-and-hold, 195
    Sampling frequency, 189-190
    Sampling jitter, 195-197
    Successive approximation, 202
    Time-interleaved, 202-203
    Track-and-hold, 194-195
    Voltage staircase waveform, 193-194
Ansys, 42
Antenna
    Active electronically-scanned array (AESA), 117, 119-121, 273
    Bandwidth, 105, 114, 118
    Beamwidth, 108-109
    Circularly polarized, 118
    Dipole, 107, 109, 112
    Directivity, 106, 109-111, 117
    Equivalent isotropic radiated power (EIRP), 112
    Feed, 106, 110, 113, 117
    Flat panel, 117, 119
    Gain, 106, 111-112, 113, 115, 117, 118
    High gain, 117
    Isotropic radiator, 106, 108, 112, 117
    Mismatch, 113, 114
    Parabolic reflector, 110-111, 116-117

Antenna (continued)
  Radiated power, 109-112
  Sidelobes, 108
Asymmetric digital subscriber line (ADSL), 273
Attenuator, 4, 78-80, 114-115, 130-131, 144-145, 154-155, 266

B

Balanced mixer, 15, 230, 232, 237, 239
Balanced amplifiers, 135-136, 181
Balun, 4, 16, 22, 63, 72-74, 181, 234, 235, 237
Bandgap, 24-25, 36, 41, 274, 275
Bandwidth efficiency, 248-250
Binary phase-shift keying (BPSK), 251-254, 257, 259
Bipolar junction transistor (BJT), 3, 18, 39, 273
Bit error rate (BER), 248, 257, 273
Bond wire, 46, 73

C

Capacitor, 4, 13, 22, 26, 34, 43-45, 79, 90, 101-102, 125-126, 130, 132, 172, 174-175, 177, 182, 195, 212, 214, 228, 234, 245
Carrier to intermodulation ratio (CIMR), 207, 273
Cascode, 236
CDMA, 260-262, 273
Ceramic, 8, 47
Channel capacity, 249-250
Characteristic impedance, 49-50, 51-54, 60, 66, 69-71, 90-91, 93-95, 97, 106, 132-133, 157-158, 179, 213
Circularly polarized antennas, 118
Coaxial cable, 118
Coaxial line, 49-50, 51, 98
Combiner, 4, 63, 67-68, 181
Commercial off-the-shelf (COTS), 121, 273
Complementary metal oxide transistor (CMOS), 9, 15, 18-19, 22, 28, 35-36, 42, 76, 78, 80, 83-84, 101, 163-164, 235, 262-263, 273
Conductance, 132
Conduction angle, 169-174

Conduction band, 24, 27, 30
Conductivity (electrical), 26, 28, 38
Conductivity (thermal), 28
Co-planar waveguide (CPW), 60-61, 67
Coupler, 60, 63-67
Coupling, 63, 64, 66, 74, 93, 95, 99, 100, 213, 246
Crystal oscillator, 10, 14, 15, 16, 210-212, 214, 229, 246

D

DC block, 125, 175, 214, 228
Dielectric resonator, 14, 98-100, 209, 212-214
Dielectric strength, 28
Digital predistortion (DPD), 21, 82, 178, 186, 273
Digital signal processor (DSP), 11, 12, 14, 81-82, 121, 180, 190, 273, 275
Digital-to-analog converters (DACs)
  Basis, 204
  Characteristics, 205-207
  Output voltage, 204-205
Diode, 4, 15-19, 23-26, 28-33, 41-42, 75-76, 130, 148, 185, 214, 217, 220-221, 226-233, 245-246, 274-275
Direct digital synthesis (DDS), 83
Directivity (antenna), 106, 109-111, 117
Discontinuities, 57, 58, 91
Dispersion, 54, 55, 56, 60, 64, 133, 138
Distributed, 134-139
Doherty PA
  Block scheme, 179
  Linearity, 181
Double-balanced mixer, 15, 237, 239
Downconversion
  Direct, 225-232
  Superheterodyne, 145
Duplexer, 4
Dynamic range, 124, 129, 191, 206, 236

E

E-band, 7, 20, 67, 68, 232, 239
Energy per bit, 247, 248, 249, 258, 259
Effective aperture area, 111, 113

Effective isotropic radiated power (EIRP), 112
Effective radiated power (ERP), 112
Electronic bandgap (EBG), 24-25, 36, 41
Electronically-scanned antenna (ESA), 117, 119-121, 273
Envelope tracking PAs, 180-181
Equalizer, 4
Equivalent circuit
  BJT, 40
  Capacitor, 44-45
  FET, 37
  HEMT, 40
  Inductor, 46-47
  Large-signal, 21, 34, 35, 39, 40, 166, 168
  Resistor, 48-49
  Small-signal, 32-40, 123, 127, 166, 229, 269, 270
  Varactor diode, 33
Equivalent noise temperature, 10, 114, 142-144, 152, 153, 155
Equivalent number of bits (ENOB), 191-193, 273
Equivalent series resistance, 44
Error vector magnitude (EVM), 167-168, 273
Exclusive-OR gate, 81, 260-261

F

Fading
  Rayleigh, 257, 260
Feedback amplifier, 216, 217
Field-effect transistors (FETs), 36-37, 273
Field programmable gate array (FPGA), 83, 84, 118, 273
FinFET, 84
Flash ADCs, 200-201
Flicker noise, 10, 33, 47, 148-150, 217
Folding ADCs, 201
Frequency bands, 6-9, 12, 87, 225
Frequency division multiple access (FDMA), 262
Frequency division multiplexing (FDM), 256, 273
Frequency multipliers
  Characteristics, 15, 220-221
  Upconversion, 225, 233

Friis' equation, 112-113, 115
Front end, 9, 105

G

Gallium
  Arsenide (GaAs), 9, 15, 17, 18, 19, 20, 22, 23, 25, 26, 28, 32, 36, 37, 38, 41, 42, 56, 75, 76, 77-79, 120, 147, 156, 159, 162, 164, 165, 182, 214, 232, 273
  Nitride (GaN), 9, 17, 23, 28, 120, 274
GaN HEMTs, 19, 20, 22, 37-40, 42, 125, 127, 136, 162, 164, 165, 177, 181, 182, 274
GaN MMIC PA, 182-183
Gates (digital), 80-81
Gilbert cell
  Bipolar technology, 16, 237-239
  Simplified circuit, 237
Grounding, 50, 60
Guided wavelength, 49-56, 59, 60, 66, 97

H

Halogen, 147
Harmonic distortion (DACs), 206-207
Heterojunction bipolar transistor (HBT), 16, 41, 274
High electron mobility transistor (HEMT)
  GaN, 19-20, 22, 37-39, 40, 42, 125, 127, 136, 162, 164-165, 177, 181-182, 274
  Large-signal model, 40
Hybrid microwave integrated circuit (HMIC), 51

I

Indium phosphide (InP), 23, 274
Inductor, 4, 13, 22, 26, 41, 43, 45-47, 90, 101, 132-133, 172, 174-175, 182, 212, 214, 228, 234
Intercept point, 184, 274
Interference, 256, 263
Intermodulation
  Distortion (IMD), 167, 183-185, 207, 224-225, 236, 273
Isolation, 4, 64, 68-71, 75, 77, 135, 225, 230-233, 256, 263

Isolator, 4, 43

K

Keysight Technologies, 42, 370, 272

L

Lange coupler, 20, 22, 47, 63, 64-68, 71, 135-136, 181, 230
Laterally diffused metal oxide semiconductor (LDMOS), 36, 274
Limiter, 4
Linearity
    Class AB PAs, 170-171
    Doherty PA, 179, 180-181
Linearization techniques
    Analog IF predistortion, 185-186, 273
    Digital predistortion, 21, 82, 178, 185-186, 273
Local oscillator (LO)
    Crystal, 209-212
    Phase noise, 215-220
    Pierce, 212
    Voltage-controlled (VCO), 214-215
Loss
    Conductor, 48
    Dielectric, 26, 28
    Insertion, 14, 71-72, 75, 77, 88-90, 95, 98, 100-101, 270
    Radiation, 48, 60, 95
    Return, 106, 124, 134, 136
Low-noise amplifiers (LNAs)
    Biasing, 126
    Cascaded block effects, 151-155
    Design for minimum noise figure, 156-159
    Gain, 143
    Noise figure (NF), 142-145
Low-noise block (LNB), 116-117
LTE-Advanced, 250
Lumped element, 12, 92-95

M

Matching network synthesis (max. gain-oriented), 113, 131-134
Matching network synthesis (min. noise-oriented), 156-159

Metal semiconductor field-effect transistor (MESFET), 36, 273-274
Metal-oxide semiconductor field-effect transistor (MOSFET), 274
Metal-insulator-metal (MIM) capacitors, 101-102, 274
Microstrip, 8, 13, 22, 50-62, 64, 66-67, 69, 70-73, 90-102, 113, 117, 119, 131-134, 137, 157, 177-178, 182, 213-214, 231, 277
Millimeter-wave, 3, 6-8, 12-13, 18-19, 36, 39, 67, 79, 84, 90, 98, 101-102, 117, 127, 129, 136, 151, 162-163, 169, 172, 181-182, 186-187, 207, 220, 232, 272, 274
Mixer
    Balanced, 230-232
    Cascode, 236
    Double-diode, 230-231
    FET-based, 233-236
    Gilbert cell, 15-16, 226, 236-239
    Image-reject, 231-232
    Transistor-based, 233-239
Modulation
    Amplitude (AM), 150, 241-245
    Digital, 247-262
    Frequency (FM), 245-247
    Index, 251, 257
Monolithic microwave integrated circuit (MMIC), 1-2, 12-13, 16-20, 22, 27-28, 34, 36, 39, 41-42, 44, 45, 46-47, 50-51, 56-57, 60-61, 64, 67-68, 71, 74, 76, 80, 95, 97, 101-102, 120, 126, 129, 131, 136-138, 146, 156-157, 159, 161-162, 181-182, 189, 231, 237, 274, 275
Multiple-input multiple-output (MIMO), 5, 118-119, 263, 274

N

NAND gate, 80-81, 83
NI AWR, 42, 66-67, 71, 95-96, 128, 134, 157, 181, 272
Noise
    Equivalent resistance, 218

Index

Factor, 142-145, 150-155, 217-218
Figure (NF), 10, 22, 41, 142-155, 225, 233, 274
Transistor, 146-151
Noise sources
    Flicker, 10, 33, 47, 148-150, 217
    Quantization error, 192-194
    Sampling distortion, 194-195
    Shot, 33, 147-148
    Thermal noise, 10, 32, 114, 146-150, 195, 215-217
Noise temperature, 10, 105, 114-115, 142-145, 152-155
Nonlinearities, 166, 183, 270-272
NOR gate, 80, 83
Numerically controlled oscillator (NCO), 191, 274

O

Orthogonal frequency-division multiplexing (OFDM), 256-257, 262, 274

P

Package, 21, 46, 118-119, 124, 157, 181, 210, 275
Parasitic, 20, 41, 44, 46, 48, 79, 93, 162, 177, 182
Peak-to-average power ratio (PAPR), 124, 167, 192, 274
Phase-locked loop (PLL), 15, 222-224, 274
Phase noise
    Amplifiers, 149-150
    Analysis, 215-220
    Leeson's formula, 215
    Oscillators, 215-220
Pinch-off voltage, 78, 148
Pipelined ADCs, 201-202
Polymer, 20, 49, 51, 96, 133, 162
Power-added efficiency (PAE), 124, 165-167, 178, 274
Power supply, 111, 126, 172, 181
Predistortion
    Analog, 273
    Digital, 21, 82, 178, 186, 273

Propagation velocity, 49-51
Pseudomorphic high electron mobility transistor (pHEMT), 19-20, 22, 26, 37-38, 42, 147, 156, 159, 162, 164-165, 182, 274

Q

Quadrature amplitude modulation (QAM)
    Demodulator and digital processing, 255
    Modulated signal representation, 255
    Symbol concepts and processing, 255
Quadrature phase-shift-keying (QPSK), 249, 251-252, 254-255, 257-259
Quality factor (Q-factor), 13, 32, 44-44-47, 59, 61, 89, 99-101, 134, 211, 214, 216, 218
Quasi-TEM, 8, 51, 54
Quantization error, 192-196
Quartz crystal
    Stabilized oscillator, 14-16, 209-212, 215, 218-219
Quiescent point (Q-point), 168-171

R

Radio-frequency integrated circuit (RFIC), 12, 16-19, 22, 27-28, 34-36, 41, 46, 48, 51, 57, 61, 73, 76, 129, 131, 146, 156, 189, 202-203, 237, 275
Radio-frequency identification (RFID), 275
Receiver, 4, 10-12, 101, 111, 114-117, 119-120, 129-131, 142, 144-146, 151, 153, 154, 192, 204, 225, 228, 231, 248, 255, 260-263, 274-275
Reflected voltage, 135
Reflection coefficient, 128, 135-136, 213-214, 270
Resistance
    Electrical, 26-27, 32, 40-41, 44-48, 50, 75, 77, 79, 136, 146-147, 172 176, 182, 193, 195-196, 213, 217-218, 220
    Noise, 218
    Thermal, 10, 32, 114, 146-150, 195, 215-217
Resistivity, 28, 45, 47, 48

Resistor, 5, 10, 22, 26-27, 34, 43, 45, 47-49, 68-69, 71, 77-79, 114, 144, 182, 200, 212, 214, 235
Resonator, 13, 14, 92-95, 98-100, 209-212, 213-214, 216
RF choke (RFC), 125, 178, 212, 228, 236
RF spectrum, 10

S

Sampling
    Clock, 191-192, 195
    Concept, 14, 189
    Distortion, 194-195
    Frequency, 190, 198-199
    Jitter, 195-197
    Theorem, 189-190
Scattering matrix (S parameters), 269-270
Selectivity, 101
Semiconductor, 1, 8-9, 15, 17-19, 23-43, 45, 48, 50-51, 75, 101-102, 125, 148, 162, 194, 206, 220, 226, 228, 273-275
Sensitivity, 64, 120, 124, 247
Sidelobes, 108
Silicon, 1, 9, 15, 17, 20, 23, 25-26, 28, 30, 32, 35-36, 39, 41, 48, 162, 262-263, 273-275
Silicon-germanium (SiGe), 15, 16, 18, 23, 25-28, 36, 42, 156, 163, 237, 275
Signal-to-noise ratio (SNR), 10, 115-116, 142, 146, 191-192, 206, 248, 249, 275
Signal, noise and distortion (SINAD), 145, 206, 275
Single-ended diode mixer, 228-230, 232
Silicon RF CMOS, 18, 35, 36, 42, 163
SONNET©, 46
Spectral efficiency, 167, 248, 250, 251, 254-257, 260
Splitter, 5, 63, 71-72, 181
Spread spectrum modulation, 260-262
Spurious signal, 183
Spur-line bandstop filter, 97-98, 231
Stability measure, 128-129, 138
Substrate integrated waveguide (SIW), 61, 100, 101
Successive-approximation ADCs, 202

Superheterodyne receiver, 145
Switch
    FET, 75-76, 77-78
    PIN, 75-76

T

Thermal
    Conductivity, 28
    Noise, 10, 32, 114, 146-150, 195, 215-217
    Resistance, 218
3G, 261
Time-division multiplexing (TDM), 275
Time-interleaved ADCs, 202-203
Transconductance, 37, 41, 168, 230, 236-237
Transition frequency, 36, 37, 40-41, 127
Transceiver, 117, 118, 161, 241, 262-263, 275
Transmitter, 4, 112, 119, 161, 262-263, 275
Transverse electromagnetic (TEM), 49, 275
Transverse magnetic (TM), 59
Travelling wave tube (TWT), 136
Tuning, 212, 216
Two-dimensional electron gas (2-DEG), 38-39, 273

U

Upconversion
    Direct, 233
    Frequency multipliers, 15, 220-221
    Mixers, 232-233

V

Varactor, 5, 15, 18, 31-33, 102, 214, 217, 275
Variable-gain amplifier (VGA), 80
Voltage-controlled oscillators (VCOs), 14, 16, 32-33, 83, 191-192, 209, 214-215, 217, 222-224, 246, 275
Voltage standing wave ratio (VSWR), 66, 124, 155

W

Waveguide, 7, 22, 61, 96, 98, 100, 136, 277
Wilkinson power divider, 12, 22, 47, 63, 67-72

X

X-parameters, 35, 269, 270-272

Artech House Microwave Library

Behavioral Modeling and Linearization of RF Power Amplifiers, John Wood

Chipless RFID Reader Architecture, Nemai Chandra Karmakar, Prasanna Kalansuriya, Randika Koswatta, and Rubayet E-Azim

Control Components Using Si, GaAs, and GaN Technologies, Inder J. Bahl

Design of Linear RF Outphasing Power Amplifiers, Xuejun Zhang, Lawrence E. Larson, and Peter M. Asbeck

Design Methodology for RF CMOS Phase Locked Loops, Carlos Quemada, Guillermo Bistué, and Iñigo Adin

Design of CMOS Operational Amplifiers, Rasoul Dehghani

Design of RF and Microwave Amplifiers and Oscillators, Second Edition, Pieter L. D. Abrie

Digital Filter Design Solutions, Jolyon M. De Freitas

Discrete Oscillator Design Linear, Nonlinear, Transient, and Noise Domains, Randall W. Rhea

Distortion in RF Power Amplifiers, Joel Vuolevi and Timo Rahkonen

Distributed Power Amplifiers for RF and Microwave Communications, Narendra Kumar and Andrei Grebennikov

Electric Circuits: A Primer, J. C. Olivier

Electronics for Microwave Backhaul, Vittorio Camarchia, Roberto Quaglia, and Marco Pirola, editors

EMPLAN: Electromagnetic Analysis of Printed Structures in Planarly Layered Media, Software and User's Manual, Noyan Kinayman and M. I. Aksun

An Engineer's Guide to Automated Testing of High-Speed Interfaces, Second Edition, José Moreira and Hubert Werkmann

Envelope Tracking Power Amplifiers for Wireless Communications, Zhancang Wang

Essentials of RF and Microwave Grounding, Eric Holzman

Frequency Measurement Technology, Ignacio Llamas-Garro, Marcos Tavares de Melo, and Jung-Mu Kim

FAST: Fast Amplifier Synthesis Tool—Software and User's Guide, Dale D. Henkes

Feedforward Linear Power Amplifiers, Nick Pothecary

Filter Synthesis Using Genesys S/Filter, Randall W. Rhea

Foundations of Oscillator Circuit Design, Guillermo Gonzalez

Frequency Synthesizers: Concept to Product, Alexander Chenakin

Fundamentals of Nonlinear Behavioral Modeling for RF and Microwave Design, John Wood and David E. Root, editors

Generalized Filter Design by Computer Optimization, Djuradj Budimir

Handbook of Dielectric and Thermal Properties of Materials at Microwave Frequencies, Vyacheslav V. Komarov

Handbook of RF, Microwave, and Millimeter-Wave Components, Leonid A. Belov, Sergey M. Smolskiy, and Victor N. Kochemasov

High-Efficiency Load Modulation Power Amplifiers for Wireless Communications, Zhancang Wang

High-Linearity RF Amplifier Design, Peter B. Kenington

High-Speed Circuit Board Signal Integrity, Second Edition, Stephen C. Thierauf

Integrated Microwave Front-Ends with Avionics Applications, Leo G. Maloratsky

Intermodulation Distortion in Microwave and Wireless Circuits, José Carlos Pedro and Nuno Borges Carvalho

Introduction to Modeling HBTs, Matthias Rudolph

An Introduction to Packet Microwave Systems and Technologies, Paolo Volpato

Introduction to RF Design Using EM Simulators, Hiroaki Kogure, Yoshie Kogure, and James C. Rautio

Introduction to RF and Microwave Passive Components, Richard Wallace and Krister Andreasson

Klystrons, Traveling Wave Tubes, Magnetrons, Crossed-Field Amplifiers, and Gyrotrons, A. S. Gilmour, Jr.

Lumped Elements for RF and Microwave Circuits, Inder Bahl

Lumped Element Quadrature Hybrids, David Andrews

Microstrip Lines and Slotlines, Third Edition, Ramesh Garg, Inder Bahl, and Maurizio Bozzi

Microwave Circuit Modeling Using Electromagnetic Field Simulation, Daniel G. Swanson, Jr. and Wolfgang J. R. Hoefer

Microwave Component Mechanics, Harri Eskelinen and Pekka Eskelinen

Microwave Differential Circuit Design Using Mixed-Mode S-Parameters, William R. Eisenstadt, Robert Stengel, and Bruce M. Thompson

Microwave Engineers' Handbook, Two Volumes, Theodore Saad, editor

Microwave Filters, Impedance-Matching Networks, and Coupling Structures, George L. Matthaei, Leo Young, and E. M. T. Jones

Microwave Material Applications: Device Miniaturization and Integration, David B. Cruickshank

Microwave Materials and Fabrication Techniques, Second Edition, Thomas S. Laverghetta

Microwave Materials for Wireless Applications, David B. Cruickshank

Microwave Mixer Technology and Applications, Bert Henderson and Edmar Camargo

Microwave Mixers, Second Edition, Stephen A. Maas

Microwave Network Design Using the Scattering Matrix, Janusz A. Dobrowolski

Microwave Radio Transmission Design Guide, Second Edition, Trevor Manning

Microwave and RF Semiconductor Control Device Modeling, Robert H. Caverly

Microwave Transmission Line Circuits, William T. Joines, W. Devereux Palmer, and Jennifer T. Bernhard

Microwaves and Wireless Simplified, Third Edition, Thomas S. Laverghetta

Modern Microwave Circuits, Noyan Kinayman and M. I. Aksun

Modern Microwave Measurements and Techniques, Second Edition, Thomas S. Laverghetta

Modern RF and Microwave Filter Design, Protap Pramanick and Prakash Bhartia

Neural Networks for RF and Microwave Design, Q. J. Zhang and K. C. Gupta

Noise in Linear and Nonlinear Circuits, Stephen A. Maas

Nonlinear Microwave and RF Circuits, Second Edition, Stephen A. Maas

On-Wafer Microwave Measurements and De-Embedding, Errikos Lourandakis

Passive RF Component Technology: Materials, Techniques, and Applications, Guoan Wang and Bo Pan, editors

Practical Analog and Digital Filter Design, Les Thede

Practical Microstrip Design and Applications, Günter Kompa

Practical Microwave Circuits, Stephen Maas

Practical RF Circuit Design for Modern Wireless Systems, Volume I: Passive Circuits and Systems, Les Besser and Rowan Gilmore

Practical RF Circuit Design for Modern Wireless Systems, Volume II: Active Circuits and Systems, Rowan Gilmore and Les Besser

Production Testing of RF and System-on-a-Chip Devices for Wireless Communications, Keith B. Schaub and Joe Kelly

Q Factor Measurements Using MATLAB®, Darko Kajfez

QMATCH: Lumped-Element Impedance Matching, Software and User's Guide, Pieter L. D. Abrie

Radio Frequency Integrated Circuit Design, Second Edition, John W. M. Rogers and Calvin Plett

Reflectionless Filters, Matthew A. Morgan

RF Bulk Acoustic Wave Filters for Communications, Ken-ya Hashimoto

RF Design Guide: Systems, Circuits, and Equations, Peter Vizmuller

RF Linear Accelerators for Medical and Industrial Applications, Samy Hanna

RF Measurements of Die and Packages, Scott A. Wartenberg

The RF and Microwave Circuit Design Handbook, Stephen A. Maas

RF and Microwave Coupled-Line Circuits, Rajesh Mongia, Inder Bahl, and Prakash Bhartia

RF and Microwave Oscillator Design, Michal Odyniec, editor

RF Power Amplifiers for Wireless Communications, Second Edition, Steve C. Cripps

RF Systems, Components, and Circuits Handbook, Ferril A. Losee

Scattering Parameters in RF and Microwave Circuit Analysis and Design,
  Janusz A. Dobrowolski

The Six-Port Technique with Microwave and Wireless Applications,
  Fadhel M. Ghannouchi and Abbas Mohammadi

Solid-State Microwave High-Power Amplifiers, Franco Sechi and Marina Bujatti

Spin Transfer Torque Based Devices, Circuits, and Memory, Brajesh Kumar Kaushik
  and Shivam Verma

Stability Analysis of Nonlinear Microwave Circuits, Almudena Suárez and
  Raymond Quéré

Substrate Noise Coupling in Analog/RF Circuits, Stephane Bronckers,
  Geert Van der Plas, Gerd Vandersteen, and Yves Rolain

System-in-Package RF Design and Applications, Michael P. Gaynor

Technologies for RF Systems, Terry Edwards

Terahertz Metrology, Mira Naftaly, editor

TRAVIS 2.0: Transmission Line Visualization Software and User's Guide, Version 2.0,
  Robert G. Kaires and Barton T. Hickman

Understanding Microwave Heating Cavities, Tse V. Chow Ting Chan and
  Howard C. Reader

Understanding Quartz Crystals and Oscillators, Ramón M. Cerda

For further information on these and other Artech House titles, including previously considered out-of-print books now available through our In-Print-Forever® (IPF®) program, contact:

Artech House Publishers
685 Canton Street
Norwood, MA 02062
Phone: 781-769-9750
Fax: 781-769-6334
e-mail: artech@artechhouse.com

Artech House Books
16 Sussex Street
London SW1V 4RW UK
Phone: +44 (0)20 7596 8750
Fax: +44 (0)20 7630 0166
e-mail: artech-uk@artechhouse.com

Find us on the World Wide Web at: www.artechhouse.com